SpringerBriefs in Food, Health, and Nutrition

Editor-in-Chief

Richard W. Hartel, University of Wisconsin—Madison, USA

Associate Editors

J. Peter Clark, Consultant to the Process Industries, USA
John W. Finley, Louisiana State University, USA
David Rodriguez-Lazaro, ITACyL, Spain
David Topping, CSIRO, Australia

For further volumes:
http://www.springer.com/series/10203

Franco Maria Ruggeri · Ilaria Di Bartolo
Fabio Ostanello · Marcello Trevisani

Hepatitis E Virus

An Emerging Zoonotic and Foodborne Pathogen

 Springer

Franco Maria Ruggeri
Ilaria Di Bartolo
Department of Veterinary Public Health
  and Food Safety
Istituto Superiore di Sanità
Rome
Italy

Fabio Ostanello
Marcello Trevisani
Department of Veterinary
  Medical Sciences
Bologna University
Ozzano Emilia (BO)
Italy

ISBN 978-1-4614-7521-7   ISBN 978-1-4614-7522-4   (eBook)
DOI 10.1007/978-1-4614-7522-4
Springer New York Heidelberg Dordrecht London

Library of Congress Control Number: 2013937595

Springer is part of Springer Science+Business Media (www.springer.com)

Contents

Chapter 1
Introduction

Hepatitis E, previously known as "enterically transmitted non-A, non-B, non-C (non-A-C) hepatitis", is an infectious viral disease with clinical and morphological features of acute hepatitis. The course of disease is mild in most affected people, except pregnant women, for whom the mortality rate can rise to 20 % (Aggarwal and Jameel 2011). The etiological agent, first identified in the early 1980s, is the hepatitis E virus (HEV) (Emerson and Purcell 2003). The disease is an important public health concern in developing countries (Southeastern and Central Asia, the Middle East, northern and western areas of Africa and America) where it is frequently epidemic, and is mainly transmitted by the fecal-oral route usually through the consumption of contaminated water or food (Emerson and Purcell 2003; Wibawa et al. 2004; Aggarwal 2011).

Industrialized countries, such as Canada, Europe, Japan and the USA were previously thought to be exempt from HEV, with a limited number of cases reported only in people who had travelled to endemic areas of the world. However, more recent studies have documented a number of sporadic cases in developed areas, including the USA and Europe, among patients who had no history of travelling to hepatitis E endemic countries (Aggarwal and Jameel 2011). Furthermore, a high anti-HEV seroprevalence (in some cases reaching 20 %) has been detected in a significant proportion of healthy individuals of non-endemic countries (Aggarwal and Jameel 2011).

Since the early 1990s, serological evidence of HEV infections and in some cases virus detection have been reported in many animal species such as rhesus monkeys, pigs, cattle, sheep, poultry, dogs, cats, goats, rabbits, mongooses, bats and rodents, both in developed and developing countries, suggesting the possibility that these species may become infected with HEV-like viruses (Huang et al. 2002; Emerson and Purcell 2003). However, for some species, such as dogs, goats and cats, the virus has never been recovered or sequenced up to now, and further studies are, therefore, needed to fully understand the significance of seropositivity in these animals. In 1997, a swine HEV strain was identified for the first time in the USA, and named swine hepatitis E virus (swHEV) (Meng et al. 1997). The

F. M. Ruggeri et al., *Hepatitis E Virus*, SpringerBriefs in Food, Health, and Nutrition, DOI: 10.1007/978-1-4614-7522-4_1, © Franco Maria Ruggeri, Ilaria Di Bartolo, Marcello Trevisani, Fabio Ostanello 2013

swine virus was genetically correlated to two human HEV strains isolated in the USA in the same period from patients affected with hepatitis E, who had not travelled to endemic areas (Meng et al. 1997). Since then, swine HEV strains have been isolated across the globe. Frequently, a strict genetic correlation between human and swine strains from the same geographic region has been observed and, during experimental infections, the possibility of cross-species transmission of swine strains to humans and of human strains to non-human primates has been demonstrated (Meng et al. 1998a; Williams et al. 2001; Matsuura et al. 2007). Furthermore, several seroepidemiological studies have reported high antibody prevalence to HEV in people working in direct contact with swine and wild boar (Drobeniuc et al. 2001; Withers et al. 2002; Carpentier et al. 2012). The first direct evidence of a possible zoonotic transmission of HEV was provided in Japan in 2003, when cases of hepatitis E were caused by the ingestion of uncooked meat or organs from pigs, wild boar or deer, a few weeks before the onset of the disease (Tei et al. 2003; Yazaki et al. 2003; Tamada et al. 2004). More recently, a case control study conducted in France using both epidemiological and virological data confirmed that 13 human cases of hepatitis E were conclusively linked to the consumption of raw "*figatellu*" pig liver sausages (Colson et al. 2010). The disease is now recognized to represent an emerging zoonosis.

Chapter 2
Etiology

2.1 Structure of the Virus and Genome Organization

In1983, the hepatitis E virus was first identified by immune electron microscopy in the feces of patients with enterically transmitted non-A-C hepatitis. Subsequent establishment of the disease in cynomolgus macaques led to cloning and sequencing of HEV in 1990 (Balayan et al. 1983; Kane et al. 1984).

The hepatitis E virus (HEV) is a small (27–34 nm), icosahedral, non-enveloped single-stranded positive-sense RNA virus.

The HEV genome is approximately 7.2 kb in length, and presents a 7-methylguanosine cap followed by three overlapping open reading frames (ORFs) and a second non-coding region of about 65–74 nucleotides with a $3'$ poly A tail. The genome length slightly varies between animal strains; shorter genomes have been detected in rat HEV in Vietnam (6927 nt) (Li et al. 2013), in avian strains (6654 nt) (Huang et al. 2004) and in recently detected bat viruses (6767 nt) (Drexler et al. 2012), although the genome organization seemed to be conserved in all these cases. ORF1 (5073–5124 nt) codes for a non-structural poly-protein of about 1,690 amino acids, which is involved in viral genome replication and viral protein processing. The poly-protein functional domains include a methyltransferase (MeT), flanked by the Y domain, a papain-like cysteine protease (PCP), flanked by the macro domain (X domain), an RNA helicase, and an RNA-dependent RNA polymerase. The N-terminal portion of ORF1 serves as a viral methyltransferase (MeT) that catalyzes the capping of both genomic and subgenomic viral RNAs (Rozanov et al. 1992). Capping of the viral RNA determines its translation and, as recently shown, it is decisive for virus defense from the innate host response, inhibiting interferon cascade activation (Pichlmair et al. 2006). The MeT is followed by the so-called Y domain, which shares significant homology with non-structural proteins of other positive-stranded RNA viruses. This is followed by the papain-like cysteine proteases (PCP) and by the formerly designated X domain, more recently renamed as the "macro domain", which may be involved in the binding of ADP-ribose and its polymeric form (Neuvonen and Ahola 2009). As for other described PCPs, also in

F. M. Ruggeri et al., *Hepatitis E Virus*, SpringerBriefs in Food, Health, and Nutrition, DOI: 10.1007/978-1-4614-7522-4_2, © Franco Maria Ruggeri, Ilaria Di Bartolo, Marcello Trevisani, Fabio Ostanello 2013

the HEV genome, the presence of a macro domain following the PCP sequence strengthens its functional homology with proteases of diverse origin, although the specific role of HEV PCP in polyprotein (pORF1) processing still remains undefined (Ahmad et al. 2011). pORF1 is characterized by the presence of a region rich in proline residues and without a predicted secondary structure, which might act as a flexible hinge within the protein. The predicted helicase domain of HEV contains a full complement of conserved helicase motifs (Karpe and Lole 2010), including the seven conserved motifs proposed to contain both the NTPase activity and an RNA binding domain. The HEV helicase possesses an RNA 5′-triphospatase activity involved in the first step of RNA capping (Karpe and Lole 2010). The C-terminal domain of pORF1 has RNA-dependent RNA polymerase (RdRp) activity (Agrawal et al. 2001); this is an essential enzyme for RNA virus replication through the synthesis of an anti-genomic RNA intermediate. The endoplasmic reticulum was identified as the site of replicase localization, and the intracellular membranes are the possible sites where RNA replication occurs (Rehman et al. 2008). ORF3 (366–369 nt) follows ORF1 and overlaps the N-terminal portion of ORF2, in a different reading frame, and encodes for a small phosphoprotein (pORF3), which is expressed at the intracellular level. The protein contains two hydrophobic and two proline rich domains; these regions contain amino acid motifs involved in signal transduction (Korkaya et al. 2001), and a PSAP motif is present and conserved in all HEV isolates, including avian HEV. pORF3 does not show homology with any other known protein; its role still remains unclear. Recent studies of the biology of HEV replication have shown that pORF3 may be involved in virus release from infected cells (Okamoto 2011), since it is associated with the cytoskeleton and is present on the virion surface (Yamada et al. 2009) (Fig. 2.1). Moreover, pORF3 down-regulates innate host responses through the reduction of the expression of acute phase proteins and promotion of the secretion of α1-microglobulins (Panda et al. 2007).

An additional three ORFs have been described in rat and bat HEV genomes, but their function remains yet unknown. No suggestive similarity of the putative gene products of the internal reading frame to any described protein domain could be detected by BLAST comparison (Johne et al. 2010a; Drexler et al. 2012).

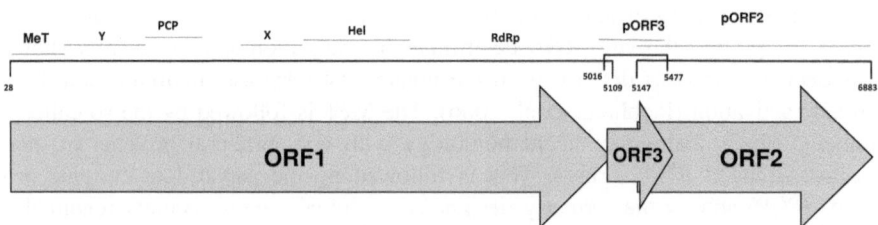

Fig. 2.1 Genomic organization of HEV, including the three ORFs. Nucleotide positions are referred to a prototype strain (Acc. No. M73218). On the *top*, functional domains are indicated: *MeT* methyltransferase; *Y* domain; *PCP* protease; *X* domain; *Hel* helicase; *RdRp* RNA dependent RNA polymerase; *pORF2* capsid protein; pORF3

The viral capsid protein encoded by ORF2 works for particle assembly, binding to host cells, and elicitation of neutralizing antibodies. pORF2 is glycosylated, and three asparagine (Asn) residues for N-linked glycosylation sites have been identified. The virus capsid is made up of subunits containing 30 homodimers of pORF2 (Yamada et al. 2009). The crystal structure of a truncated recombinant pORF2 protein has been obtained, but the real size of the protein in mature virions remains unknown (Yamashita et al. 2009). Among four major mammalian HEV genotypes, sequence identity among the amino acid residues of the capsid protein was over 85 %, and many amino acid divergences were found in the N-terminal 111 residues. The N-terminal region of the HEV capsid protein is most likely to represent the shell domain, whereas the C-terminal region of pORF2 is more variable and is considered to be the protruding domain of the HEV capsid protein (Li et al. 2009). The initial contact with host cells in order to initiate viral infection is believed to occur through these protrusions (Pichlmair et al. 2006). Expression of a truncated capsid protein lacking the first 111 amino acids and/or the C-terminal 59 amino acids in insect cells by the baculovirus expression system resulted in self-assembly of the capsid protein and in the production of two types of HEV-like particle (HEV-VLP) with different diameters (Li et al. 1997; Caprioli et al. 2005; Xing et al. 2011), corresponding to different proteolytic cleavages. As demonstrated by protein expression in the baculovirus insect cell expression system, the minimum requirement for assembly was inclusion of amino acid residues 126–601 (Li et al. 2005a). The N-terminal domain followed by the signal sequence (residues 28–101) is an arginine-rich domain resembling the RNA-binding domain of the coat proteins of tombusviruses. The capsid protein binding to Huh-7 liver cells has been studied, and it appeared to be mediated by heparin sulfate proteoglycans (HSPGs), specifically syndecans, as demonstrated using the baculovirus expressed pORF2, assembled in VLPs (Kalia et al. 2009).

Recent studies on HEV particles have provided useful information about the HEV life cycle as well as for the possible development of monovalent or polyvalent vaccines. Indeed, the recombinant HEV capsid protein is currently undergoing clinical trials as a vaccine candidate (Zhu et al. 2010; Zhao et al. 2012), and genotype 2 HEV VLPs have been proposed as a useful carrier for foreign DNA (Takamura et al. 2004) or epitopes into mucosal epithelial cells (Niikura et al. 2002). However, several knowledge gaps still remain concerning the structure of HEV capsid, such as what role pORF3 may possibly have in the virion architecture and function. Further studies will be needed to answer these questions.

2.2 Taxonomy and Nomenclature

Because of limitations in allowing it to grow reproducibly and efficiently in vitro, HEV classification has been mainly based on the analysis of the viral RNA by sequencing and phylogenetic techniques (Korkaya et al. 2001).

Table 2.1 Genotypes and host range of the hepatitis E viruses. Adapted from Meng et al. (2011)

HEV strains	Natural host
Mammalian HEV	
Genotype 1	Humans
Genotype 2	Humans
Genotype 3	Humans, domestic pigs, wild boar, deer, mongooses, rabbits
Genotype 4	Humans, domestic pigs, wild boar
Novel unclassified genotype, Rat HEV	Rats
Novel unclassified genotype, Boar HEV	Wild boar in Japan
Novel unclassified genotype, Bat HEV	Bats
Novel unclassified genotype, Ferret HEV	Ferrets
Avian HEV	
Genotype 1	Chickens
Genotype 2	Chickens
Genotype 3	Chickens
Genotype ?	Chickens (Hungary)
Trout HEV	
Genotype ?	Cutthroat trout (USA)

HEV was initially classified within the *Caliciviridae* family, but the increasing numbers of sequences collected afterwards have clearly unmasked significant differences with other caliciviruses, and since 2004 HEV has been classified as a new genus called Hepevirus in the family of *Hepeviridae* (Emerson and Purcell 2003). HEV strains detected in humans and other mammalian species represent the major genus of the *Hepeviridae* (Table 2.1). Although avian HEV strains share only 50–60 % nucleotide identity with mammalian HEV strains (Meng 2010a), specific antibodies are able to cross-react with the capsid protein of both groups of viruses, demonstrating the presence of common epitopes (Haqshenas et al. 2001). Nonetheless, avian HEV strains have never been associated with cases of infection in human beings (Kamar et al. 2012), causing hepatitis-splenomegaly syndrome (HS) only in chickens (Haqshenas et al. 2001). Consequently, it has been proposed to assign avian HEV to a separate genus, consisting of at least three different genotypes (Bilic et al. 2009; Marek et al. 2010).

Hepeviridae includes four genotypes of mammalian HEV, which primarily infect humans, domestic pigs, wild boar, deer, and rabbits (Meng et al. 2012). However, genetically distant HEV strains have more recently been identified in the rat (Johne et al. 2010b), ferrets (Raj et al. 2012), wild boar (Takahashi et al. 2011), bats (Drexler et al. 2012), and cutthroat trout (*Oncorhynchus clarkii*) (Batts et al. 2011), suggesting that the *Hepeviridae* family classification should be reviewed. A proposed revision includes introduction of separate clades: one genus would comprise human HEV genotypes and closely related animal viruses, while the others would include viruses from chiropteran (bat), rodent (rat), and avian (chicken) hosts (Drexler et al. 2012). The "cutthroat" hepevirus is genetically the most

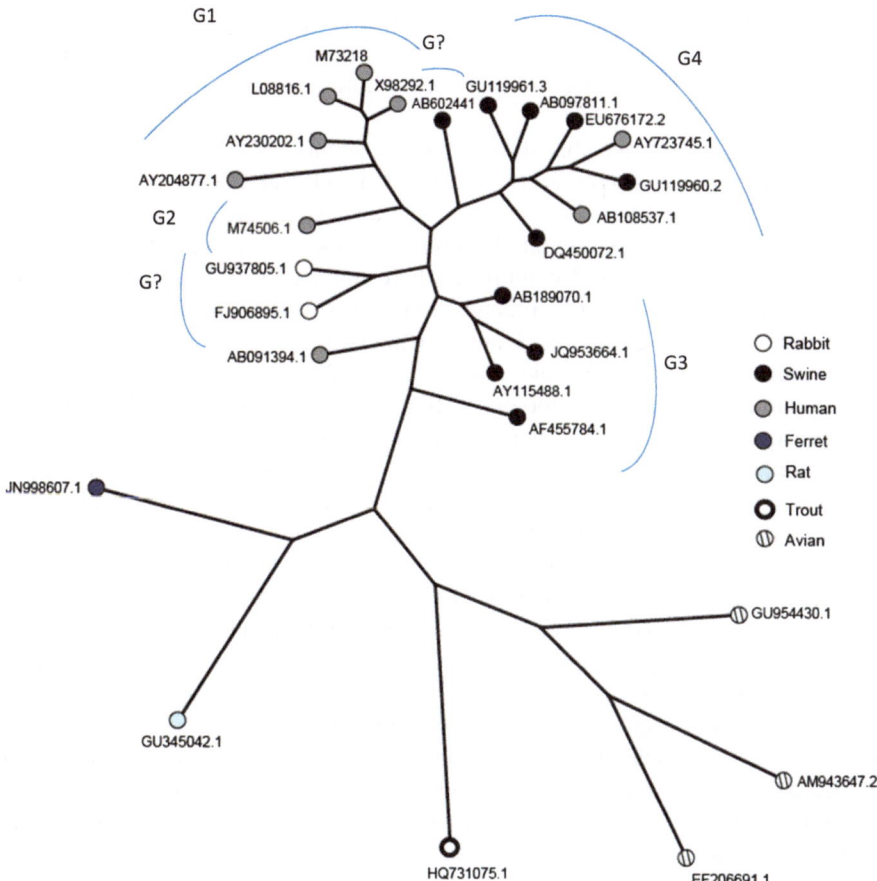

Fig. 2.2 Phylogenetic tree illustrating different genotypes of hepatitis E. The tree is based on full-length sequences of HEV strains of either human or animal origin. The GenBank accession number of each strain used in the tree is indicated

distant animal strain, and it might correspond to a separate taxonomic unit of a higher rank, e.g., a subfamily.

Based on the sequence comparisons of the HEV genome currently available, strains are classified in genotypes and sub-types. The four major genotypes identified in mammalians to date include (Fig. 2.2 and Table 2.1): genotype 1 (Burmese-like Asian strains), genotype 2 (a Mexican strain and some African strains), genotype 3 (strains from humans and animals, worldwide), and genotype 4 (strains from human cases and animal strains, in Asia and Europe) (Meng 2011). Avian HEV, genetically distant from mammalian strains, has been classified separately and comprises at least three different genotypes (Marek et al. 2010), since a putative novel avian HEV genotype has been identified but not yet classified

(Banyai et al. 2012). However, with the recent identification of genetically distinct HEV strains from rats (Johne et al. 2010a), ferrets (Raj et al. 2012), and wild boar (Takahashi et al. 2011), new previously unrecognized genotypes have been proposed.

Despite the knowledge that different HEV genotypes occur, the virus seems otherwise to exist as a single serotype (Aggarwal and Naik 2009).

Most human infections that occur in Asia and Africa are caused by genotype 1, whereas genotype 2 is commonly found in Mexico and West Africa (Nigeria and Chad). In industrialized countries, where until a few years ago hepatitis E was considered non-endemic, autochthonous cases appear to be related to HEV strains belonging to genotypes 3 and 4 (Emerson and Purcell 2003; Okamoto 2007; Kamar et al. 2012; Scobie and Dalton 2013), that are still considered the only zoonotic genotypes. Genotype 3 HEV was first identified in human cases locally acquired in the USA, when the two human strains, called US-1 and US-2, showed only 74–75 % nucleotide identity with genotypes 1 and 2, being classified separately (Meng et al. 1997). Since then, genotype 3 has been detected throughout the entire world, associated with sporadic cases and small outbreaks in North America, Europe, Japan and New Zealand (Tsang et al. 2000; Mansuy et al. 2004; Dalton et al. 2007a, b). This genotype is also commonly detected in animals, and a strict genetic correlation has been observed between human and animal strains circulating in the same geographical area. The first animal strain of HEV was identified in swine in the USA in 1997. The virus was shown to belong to genotype 3, and presented a high identity with some human strains (Meng et al. 1997). In particular, the virus shared a 92 % nucleotide identity in ORF2 with the two HEV genotype 3 autochthonous human strains (US-1 and US-2) detected in the same period in the USA. Given the strict genetic correlation, the two viruses were classified in the same genotype 3, and since then pigs have been considered a reservoir of HEV (Meng et al. 1997).

Recently, a study conducted in Japan identified a new potential reservoir of genotype 3 HEV virus in the mongoose. As revealed by phylogenetic analysis, the HEV RNAs detected belonged to genotype 3 and were classified into two groups, one of which contains sequences very similar to mongoose HEV previously detected in the same area as well as to HEV identified in a pig (Nidaira et al. 2012).

Genotype 4, the other genotype transmitted by the zoonotic route, is indigenous to Asia, where it has been recovered from both pigs and humans (Wang et al. 2012). Since its first report in China in 1999, genotype 4 has been increasingly described as being endemic in pigs and as the cause of sporadic cases of hepatitis E in humans and infection in the swine in China and Japan, and more recently in Europe (Hakze-van der Honing et al. 2011; Colson et al. 2012; Garbuglia et al. 2013). The question for the recent and increasingly frequent detection of genotype 4 in Europe is of concern for public health, and raises the question whether genotype 4 was somehow introduced into domestic pigs and may be expected to spread further on farms or whether the problem is to be correlated to importation of pig meat of Asian origin into Europe (Colson et al. 2012).

The recent availability of increasing numbers of HEV sequences emphasizes the genome diversity among HEV isolates. According to the nucleotide identity, the four genotypes were further subdivided into subtypes. HEV strains belonging to genotype 1 are more conserved and can be further classified into five subtypes (1a–1e). Genotype 2 sequences are classified into two subtypes (2a and 2b). Genotypes 3 and 4 are extremely diverse and are divided into ten (3a–3j) and seven subtypes (4a–4g), respectively. Thus, in total, at least 24 subtypes of HEV exist in nature (Lu et al. 2006). The diversity of genotypes 3 and 4 appears related to their zoonotic origin from a variety of animals in different parts of the world, whereas the relative conservation of genotypes 1 and 2 is consistent with their primary circulation in humans and a less frequent isolation from animals (Pavio et al. 2010).

Nevertheless, several studies have supported the existence of intra-host quasispecies in both humans (Grandadam et al. 2004) and swine (Bouquet et al. 2012b). A recent study conducted using next generation sequencing (NGS) confirmed the existence of intra-host quasispecies in experimentally infected swine, underlining that the range of quasispecies diversity is lower than for other human viruses, but is in line with other zoonotic viruses (Bouquet et al. 2012b). In human patients undergoing solid-organ transplantation with chronic HEV infection, the quasispecies diversification seems to be related to rapid development and progression of liver fibrosis, since patients had lower quasispecies diversification during the first year than had patients without liver fibrosis progression (Lhomme et al. 2012).

2.3 Viral Replication Cycle

Because cell culture and small animal models to investigate HEV infection was developed only recently, the viral replication and regulation processes involved are poorly understood, being mainly based on HEV genome analysis and homology with other positive-stranded RNA viruses. Nevertheless, the recent identification of various cell lines permissive to HEV replication (hepatic cell lines HuH7, PCL/PRF/5, HepG2; lung carcinoma cell lines A549; human hepatoma-derived cell line, HepaRG; porcine embryonic stem cell-derived cell line, PICM-19) (Okamoto 2011; Rogee et al. 2012) has significantly contributed to clarify some essential steps of the replicative cycle. Although no cellular receptors have been identified up to now, a role of heparin sulfate proteoglycans (HSPGs) in mediating virus attachment has been demonstrated (Kalia et al. 2009). The mechanisms by which the virus is released into the cytoplasm are unknown. Genomic RNA appears to be translated directly into the ORF1 polyprotein, even though it is unclear whether the polyprotein is or is not processed into individual functional units. The RdRp mediates the replication of the positive-sense genomic RNA into negative-sense RNA transcripts, that serve as a template for the synthesis of the full genome, and into the single 2.2 kb subgenomic RNA including the overlapping ORF2 and ORF3 that are translated in the capsid protein pORF2 and in pORF3. pORF2 is assembled in a capsid particle where the RNA genome is packaged to construct the newly formed virion. Although the role of pORF3 remains largely unclear, this protein mediates virus budding likely based on the interaction of lipid-associated virions with plasma membranes or endomembranes (Ahmad et al. 2011). Binding of the cellular TSG101 (tumor susceptibility gene 101) to pORF3 through amino acid PSAP motif (i.e., amino acids proline, serine, alanine, and proline) has been demonstrated (Surjit et al. 2006). TSG101 has been identified as the critical protein for budding of enveloped viruses, such as the human immunodeficiency virus type-1 (HIV) from the plasma membrane (Martin-Serrano et al. 2001). It is likely that pORF3 mediates virus budding by recruiting the TSG101 (Okamoto 2011).

Chapter 3
Epidemiology of the Human HEV Infection

Hepatitis E virus (HEV) infection is mainly present in developing tropical and sub-tropical countries of most of Asia, North Africa, the Middle East, and Central and South America (Emerson and Purcell 2003; Panda et al. 2007) (Fig. 3.1).

In developed countries, hepatitis E was considered a disease strictly associated with travelling through endemic countries. However, in the last decade sporadic cases or small series of cases in subjects without history of travel abroad have been recorded in the USA, Europe (including the United Kingdom, France, the Netherlands, Austria, Spain, Italy, and Greece), and in developed countries of Asia–Pacific (Japan, Taiwan, Hong Kong, Australia), suggesting the presence of autochthonous reservoirs of hepatitis E virus in these areas.

In highly endemic areas, HEV strains belonging to genotypes 1 and 2 are responsible for most endemic and epidemic cases of hepatitis E. Infection is usually transmitted among humans by the fecal-oral route, and it has proven to cause large outbreaks, particularly when associated with the consumption of contaminated water.

The genotype 1 of HEV is mainly associated with infections in Asia and Africa (Fig. 3.2), whereas genotype 2 is represented by the prototype sequence from an epidemic in Mexico, 1986, and new variants were recently identified from endemic cases in some African countries. There is no known animal reservoir for HEV genotypes 1 and 2 (Scobie and Dalton 2013) except for a genotype 1 reported in horses (Saad et al. 2007)

By contrast, in non-endemic areas infection with HEV normally causes sporadic cases or small outbreaks, and other than in imported cases, it seems to be at least partially associated with zoonotic transmission (Pavio et al. 2010). The autochthonous cases in these areas, corresponding to industrialized countries, are related to genotype 3 and 4 strains (Fig. 3.2). Of these, HEV genotype 3 is more widely distributed globally, from the USA through European countries, to China and Japan. Whereas genotype 3 HEV infects a high number of animal species (pigs, deer, wild boar, mongooses, rodents, and others), genotype 4 seems to affect only humans and pigs in East Asia, apart from recently reported infection in both swine and humans in Europe (Hakze-van der Honing et al. 2011; Garbuglia et al. 2013).

F. M. Ruggeri et al., *Hepatitis E Virus*, SpringerBriefs in Food, Health, and Nutrition, DOI: 10.1007/978-1-4614-7522-4_3, © Franco Maria Ruggeri, Ilaria Di Bartolo, Marcello Trevisani, Fabio Ostanello 2013

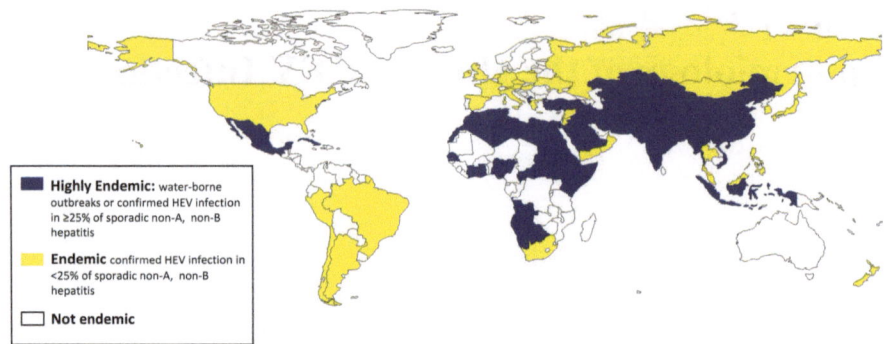

Fig. 3.1 Worldwide distribution of HEV infection in non-endemic countries, in high-endemic regions or areas, corresponding to developing countries where waterborne outbreaks are mainly described, and in areas with low endemicity, corresponding to industrialized countries

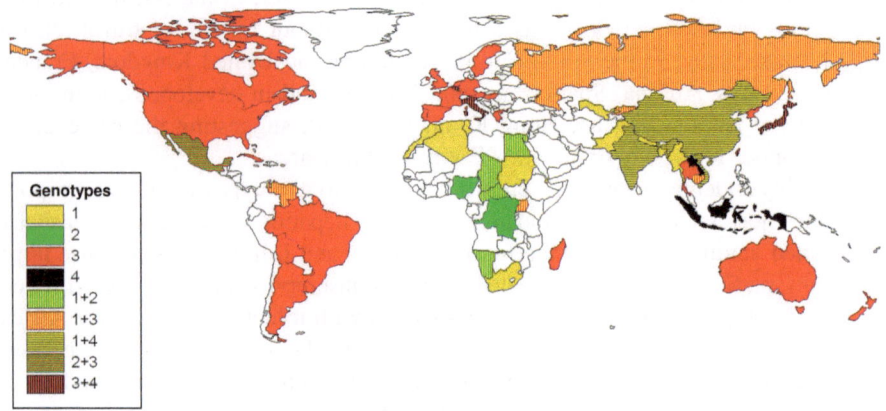

Fig. 3.2 Distribution of HEV genotypes worldwide. The *colors* used for each country represent the predominant HEV genotypes of human and animal isolates (mostly pigs), respectively, from that country

3.1 Epidemiology of Human Infection in Regions with High Disease Endemicity

In these areas, the disease usually occurs with epidemic outbreaks that affect large parts of the population, and are often separated by intervals of a few years. These outbreaks have been observed in the Indian subcontinent, China, Southeast and Central Asia, the Middle East, and the northern and western parts of Africa. In North America (Mexico), two small outbreaks took place between 1986 and 1987, but no further outbreaks have since been reported (Aggarwal and Jameel

2011). Outbreaks are often large, and some reports account for outbreaks affecting thousands of people, particularly in China, India, Somalia, and Uganda (Zhuang et al. 1991; Naik et al. 1992; Bile et al. 1994; Teshale et al. 2010). Most reported outbreaks have been related to the consumption of drinking water which had been contaminated with human feces, and infection had clearly been transmitted through the classical fecal-oral route. The time course of outbreaks in highly endemic countries varies from a few weeks to more than 1 year, and outbreaks with a more extended duration are likely to be related to a persisting source of water contamination, such as sewage systems intersecting surface water reservoirs, lakes, rivers, or clean water pipelines (Kamar et al. 2012). Particularly, the large outbreaks frequently follow heavy rainfall and floods, which may abruptly determine the discharge of human excreta into the sources of drinking water. Nonetheless, some outbreaks have occurred in hot and dry months, possibly as a result of diminished water flows in rivers that may have led to an increased concentration of fecal contaminants. In Southeast Asia, recurrent epidemics have been shown to be associated with disposal of human excreta into rivers and subsequent use of water from the same river for drinking, cooking, and personal hygiene. Outbreaks of hepatitis E have occurred in underdeveloped urban areas with leaky water pipes passing through the soil, which was found to be contaminated by sewage. Intermittent water supply in these areas leads to negative pressure in pipes during periods with no flow, allowing inward suction of contaminants (Sailaja et al. 2009). Although the dissemination of HEV infection through contamination of food may be possible, few outbreaks related to food-borne transmission have been reported from hepatitis E endemic areas. Less frequent routes of transmission include contaminated food and transfusion of infected blood products (Aggarwal and Jameel 2011). Direct person-to-person transmission seems to be uncommon; however, in a large outbreak recently reported in Uganda household factors (e.g., hygiene measures, storage of drinking water in large-mouthed containers) have been considered to be significant for increasing transmission of infection (Howard et al. 2010).

Overall attack rates during hepatitis E outbreaks have ranged from 1 to 15 % (Aggarwal and Jameel 2011). The rates are higher among young adults (3–30 %), but the reason for this is not clear. Lower attack rates among children may be partially related to a higher proportion of asymptomatic infections at younger ages and not only to an authentic scarceness of infection in young individuals.

However, in most endemic areas, seroprevalence in children below 10 years of age is approximately 5 %, whereas the ratio rises to 10–40 % among adults over the age of 25 years (Emerson and Purcell 2003).

The disease can be long lasting. Men are clinically infected two to five times more frequently than women; this may be due either to their greater risk of exposure to contaminated water, or may indicate that men develop a symptomatic infection more frequently. During hepatitis E outbreaks, pregnant women have a higher disease attack rate, and are more likely to develop fulminant hepatic failure and die. Mortality rates reach 10–25 % in pregnant women (Boccia et al. 2006). However, once acute liver failure appears, the death rate is not different between pregnant women with hepatitis E and those with severe liver injury of other causes. Immunological factors or hormones may be responsible for this different progression of the disease in pregnant women. In disease endemic areas, HEV infection accounts for a large proportion of cases of acute sporadic hepatitis. These latter patients share similar age distribution, severity and duration of illness, predisposition to worsened prognosis in the case of pregnant women, and absence of chronic sequelae with those involved in HEV outbreaks (Aggarwal and Jameel 2011).

In India, HEV infection is the most common cause of acute sporadic hepatitis, accounting for up to 70 % of all cases among adults. The route of transmission of infection in most patients with sporadic hepatitis E is unclear (Aggarwal and Naik 2009). However, given the low hygienic conditions in underdeveloped areas of India and other Asian countries, the major sources of infection are most likely water and food contaminated with human feces. Asymptomatic infections were estimated to exceed the number of symptomatic cases, by two to four times. HEV genomic sequences were detected in nearly 40 % of sewage specimens collected in a large Indian city throughout the year which clearly indicates the ubiquitous circulation of HEV through the population, even when no disease outbreak or symptomatic cases were recorded.

Unlike several other infections with fecal-oral transmission, person-to-person transmission of HEV from epidemic or sporadic cases is considered to be uncommon (Aggarwal and Jameel 2011). The exact reason for this is unknown, although differences in the minimal infectious dose able to cause overt disease and burden of virus shed in the stools by infected patients, or the environmental resistance of different viruses may play a major role (Emerson and Purcell 2003). The secondary attack rate observed among household contact of patients involved in hepatitis E outbreaks is normally much lower (0.7–2.2 %) than observed among susceptible household contacts of cases for hepatitis A (50–75 %).

The short time lapse normally observed between onset of symptoms in distinct family members during multi-case houschold outbreaks also supports a reduced person-to-person transmission, rather indicating a common primary source of infection (e.g., water, or food) (Aggarwal and Jameel 2011).

Maternal–fetal transmission of HEV infection has been reported, and the occurrence of HEV viremia among healthy blood donors and transmission of infection to transfusion recipients have been documented in regions endemic for hepatitis E (Haim-Boukobza et al. 2012). However, the impact of parenteral and blood-borne transmission to the overall disease burden remains uncertain.

3.2 Epidemiology of Human Infection in Regions with Low Endemicity

HEV infection is now thought to be endemic also in many industrialized countries. In Europe, the USA, and Japan, increasing reports of sporadic cases have been made in patients who had never travelled to foreign countries. Strains isolated in these cases were demonstrated to be genetically different from strains isolated in other regions, leading to the hypothesis that they were related to viruses endemic to the specific country (van der Poel et al. 2001; Clemente-Casares et al. 2003; Takahashi et al. 2003). Autochthonous cases of hepatitis E have been detected in all developed countries where they have been searched for, with the exception of Finland (Kantala et al. 2009). Several studies have also reported high HEV sero-prevalence rates (5–20 %) among healthy individuals in industrialized countries, suggesting a wide spread infection, which most likely occurs at a subclinical level (Emerson and Purcell 2003; Mansuy et al. 2011). The actual percentage of sub-jects seropositive to HEV might be proven to be even higher, when further studies are carried out using the more sensitive tests recently developed for anti-HEV anti-body detection. Accordingly, the seroprevalence among blood donors in Toulose (France) was shown to rise from 16 to 52 % using a modern assay with higher sensitivity than previously available assays (Kamar et al. 2012), that may obvi-ously apply to other regions investigated in the past. Since large part of infections occurring in developed countries seem to be asymptomatic, this observation could explain the high seroprevalence rates opposed to the low number of cases reported. Differently than genotypes 1 and 2, infections caused by genotypes 3 and 4 HEV strains appear to cause clinical hepatitis in middle-aged subjects and the elderly. This unusual and peculiar demography also remains unexplained, since exposure to HEV is independent of age and sex, and suggests the presence of host risk fac-tors that may be crucial for clinical manifestation of the disease. Moreover, the high mortality associated with pregnancy and genotype 1 HEV infections has not been reported with either genotypes 3 or 4 strains. Nonetheless, these latter have never been reported in such large outbreaks as genotype 1 HEV strains have, mak-ing absolute comparisons difficult with regard to pregnant women infected. In developed countries, the disease appears only as sporadic cases or small outbreaks, and a range of severity of illness has been described, from asymptomatic to acute or subacute liver failure.

In Italy, as an example of a modern country with high hygienic conditions, HEV infection is thought to account for approximately 5–10 % of cases of acute non-A–C viral hepatitis (Zanetti and Dawson 1994), which would lead to esti-mates of just few hundred cases per year. As in other developed areas, most cases occurring in Italy are associated with travel to endemic areas being reported on return from regions traditionally considered endemic, and the first identification of an autochthonous HEV dates back to 1999, when a genotype 3 virus similar to American strains was identified in a patient who had neither travelled to nor had contact with individuals associated with endemic areas (Schlauder et al. 1999). A recent detailed paper has attributed to HEV the etiology of 134 of 651 (20.6 %) cases of non-A–C hepatitis hospitalized during the period 1994 through 2009 in northern Italy (Romano et al. 2011). Of these patients, 22 (16.4 %) were estab-lished as being autochthonous, whereas the other 112 were imported or associ-ated with travelers from Asia or Africa. Although molecular typing of HEV strains was possible for only five and 39 cases, respectively, it is quite remarkable that only the five indigenous strains were genotype 3 HEV whereas the imported cases were all related with genotype 1 strains. Similar findings have been found in other studies throughout Europe (Kamar et al. 2012). More recently, rare human cases linked to genotype 4 HEV strains, normally endemic in Asia in both humans and pigs (Howard et al. 2010; Wang et al. 2012) have also been reported in Europe (Wichmann et al. 2008; Tesse et al. 2012; Garbuglia et al. 2013), but it was not concluded whether any of these cases may have originated via zoonotic or food-borne transmission. Nonetheless, these findings make it obvious to ask whether genotype 4 may become widespread in Europe, as is presently genotype 3 HEV.

3.3 Routes of Transmission

The actual modes of HEV transmission in sporadic cases are still not completely understood; in addition to the ingestion of contaminated water and food, and person-to-person transmission, vertical transmission of virus from mother to infant is known to occur, while there is no evidence of sexual transmission (Kamar et al. 2012; Scobie and Dalton 2013). The possibility of HEV transmission by transfusion of blood or blood products has been documented, but its significance is still not clear (Khuroo et al. 2004; Matsubayashi et al. 2008; Haim-Boukobza et al. 2012). Zoonotic transmission has been assessed, although the extent to which this route impacts with spreading of infection is uncertain. Risk factors can include the direct or indirect contact with infected material from affected animals (professional categories such as veterinary surgeons, farmers, slaughterhouse workers, people assigned to the care of the animals may be at risk), the ingestion of food directly or indirectly contaminated (water, plants, meat products, shellfish), and parenteral transmission (Meng 2003) (Fig. 3.3).

3.3.1 Waterborne Transmission

The knowledge that surface and drinking water are a major vehicle by which HEV can massively diffuse through the population is well consolidated since four water-related epidemics occurred in Kashmir between 1978 and 1982, causing over 50,000 cases of acute liver disease and approximately 1,700 deaths (Khuroo 1991). Reports of water-related hepatitis E have concerned other countries with low sanitary conditions, water quality, and overcrowding, sometimes also in relation with political instability and war situations (Rab et al. 1997; Guthmann et al. 2006; Guerrero-Latorre et al. 2011). Heavy rains and flooding, fecal contamination of surface water reservoirs, wells, and rivers, together with unfitting water pipelines and sewage disposal networks, have all been involved in hepatitis E epidemics. In Southeast Asia, occurrence of hepatitis E outbreaks appeared to be strictly related to the use of river water for drinking, cooking, and personal washing (Razonable et al. 2011). Epidemiological evidence,

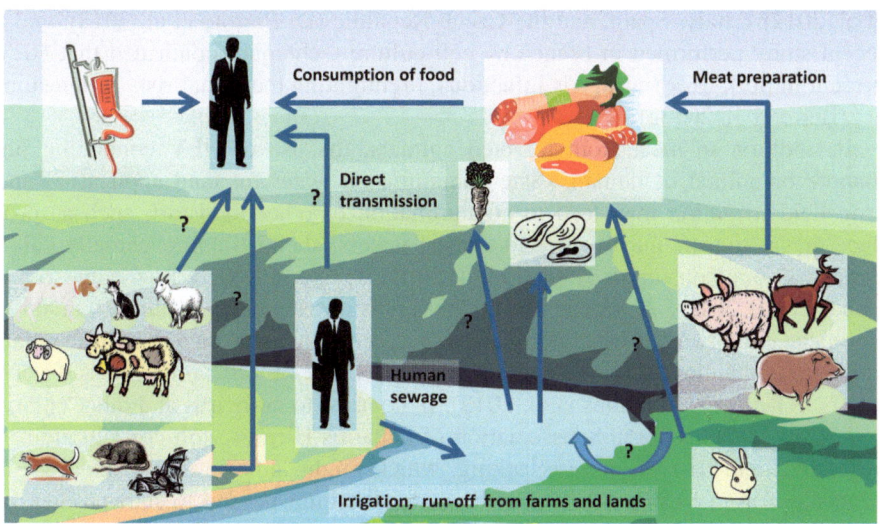

Fig. 3.3 Confirmed and potential (indicated by *question marks*) transmission routes of HEV. Genotypes 1 and 2 are waterborne only, with possible human-to-human transmission, including vertical transmission

therefore, strongly suggests that HEV can persist in environmental water, but no systematic study has been performed to define the kinetics of virus infectivity decay in the environment.

3.3.2   Foodborne Transmission

There is already some clear evidence that links onset of hepatitis E to the consumption of contaminated food items, resulting in either sporadic cases or epidemic outbreaks. The first reports regarded endemic Asian countries. In Japan, analysis of risk factors and molecular characterization of HEV from 10 patients with fulminant hepatitis E showed that the patients had eaten grilled or undercooked pig liver 2–8 weeks before onset. Some of the HEV RNA sequences found in clinical specimens were identical or similar to HEV detected in packaged pig liver sold in the market or farm swine samples (Yazaki et al. 2003). This study also showed that approximately 2 % of pig livers sold at retail in the area contained detectable viral RNA. Further observations confirming the association between pig liver or grilled pork consumption, wild boar, or deer meat, and hepatitis E were reported in the next few years in Japan (Matsuda et al. 2003, 2005; Takahashi et al. 2004; Li et al. 2005b), and other countries (Gessoni and Manoni 1996; Dalton et al. 2007b; Colson et al. 2010; Widen et al. 2011; Bouquet et al. 2012b).

Presence of HEV in food, primarily pig liver or pork, was confirmed in the USA (Feagins et al. 2007a), the Netherlands (Bouwknegt et al. 2007), UK (Berto et al. 2012b), Italy, Spain, and the Czech Republic (Di Bartolo et al. 2012), and a recent study performed in France by cell culture techniques confirmed that HEV present in pork liver sausage is infectious, highlighting the actual risk for consumers (Berto et al. 2013a).

In addition to meat from infected animals, the use of HEV containing pig manure or animal or human waste contaminated water for land application and crop field irrigation may lead to contamination of other foodstuff, such as produce or, by run off into rivers and coastal waters, shellfish and eventually cause disease among consumers (Renou et al. 2008; Song et al. 2010; Razonable 2011; Halac et al. 2012). A recent multi-laboratory investigation conducted in three European countries demonstrated that leafy green vegetables intended for the market were contaminated with HEV, finding that 3.4 % of 146 samples tested positive by RT-qPCR (Kokkinos et al. 2012). In contrast to norovirus and other enteric pathogens, data on possible association of hepatitis E cases with consumption of vegetables and berry fruits are lacking, which might indicate that residual HEV concentration in fresh produce may be under the limit needed to start productive infection in humans.

3.3.3 Professional Exposure Transmission

Human populations with occupational exposure to environmental sources of domestic animal wastes and wild animals (farmers, veterinarians, slaughterhouse personnel) have been shown to present higher anti-HEV serum antibody rates than normal blood donors or normal citizens in several studies particularly associated with the pig industry. For instance, in the USA 26 % of veterinarians were found to be seropositive to HEV compared to 18 % of regular blood donors (Meng et al. 2002), and HEV antibody prevalence was reported to be 4.5 times higher in subjects exposed to contact with pigs than in normal people (Withers et al. 2002). But similar data are available for several countries worldwide (Karetnyi et al. 1999; Meng et al. 2002; Chang et al. 2009; Masia et al. 2009; Geng et al. 2011b).

A recent investigation on forestry workers in Germany has reported a significantly higher seroprevalence against a recombinant genotype 3 HEV capsid protein fragment that reasonably suggests workers' exposure during their activity to virus probably shed by wild animals (Dremsek et al. 2012). Interestingly, a few among the forestry workers' sera also gave a strong reaction against a recombinant capsid protein from the HEV strain recently detected in rats. It is unclear how humans may have come in contact with this novel virus in order to elicit an immune response (Dremsek et al. 2012). Infectious swine hepatitis E virus was demonstrated to be present in pig manure storage facilities in farms in USA, but evidence that this may culminate into contamination of surrounding waters, although reasonable, was lacking in this study (Kasorndorkbua et al. 2005).

Conversely, both swine and human HEV strains have been reported to be present and infectious in raw sewage water in several countries (Jothikumar et al.

1993; Pina et al. 1998, 2000; Clemente-Casares et al. 2003; Ippagunta et al. 2007). In particular, sewage workers were shown to have a significantly higher anti-HEV seroprevalence than normal individuals in India, which increased with the numbers of years spent in that job (Vaidya et al. 2003), highlighting a specific occupational risk.

3.3.4 Other Routes of Transmission

As mentioned before, there is no clear demonstration sustaining that direct viral transmission of HEV from person to person is indeed an efficient method for transmitting infection. However, virus spread from infected individuals by fecal shedding in the environment inside confined areas or via contaminated fomites might play a role, especially in conditions of promiscuity and overcrowding. This might explain the difference observed in anti-HEV seroprevalence between northern and southern regions of Italy (Zanetti and Dawson 1994). In fact, the latter routinely face high immigration fluxes from countries where hepatitis E is endemic (Cacopardo et al. 1997; Scotto et al. 2013), and immigrants reside in these areas for several weeks interacting with the local population. However, the habit of eating raw shellfish common in southern Italy could also be considered an additional risk factor, possibly also favored by HEV discharge by infected immigrants into sewage and by following coastal seawater pollution.

A marked gradient of anti-HEV seroprevalence was also shown from north to south France, but in this case the risk factors seemed to be several, including personal water supplies, fresh seafood consumption, and possession of pet pigs (Renou et al. 2008). Nonetheless, HEV circulates largely among pigs in southern France where the habit of consuming a particular fresh liver sausage (*figatellu*) is particularly marked (Colson et al. 2010), thus particular food eating habits might have at least partially contributed to the observed epidemiological differences.

A mix of transmission modes acting simultaneously may have fed the unique outbreaks involving passengers in a ship returning from a world cruise in 2008, causing overt hepatitis E with jaundice in four patients and IgM seroconversion in 4 % of the 789 subjects tested (Said et al. 2009). Overall, 25 % of passengers showed anti-HEV IgM and/or IgG, indicating both recent and past infections. The virus detected in patients was a single genotype 3 HEV similar to strains circulating in Europe, suggesting a common source of infection, and seafood consumption was a risk factor. However, from the experience built from norovirus outbreaks aboard cruise ships, it cannot be excluded that virus transmission was favored by other routes, such as water or environmental contamination within common areas.

Differently than other systemic viral infections (such as HIV, HBV and HCV), diverging data exist on the association between HEV transmission and injection drug use (IDU). Specific seroprevalence ranged from 2 to over 60 % in different countries, although significant differences between IDU subjects and healthy blood donors have not been observed in all cases (Gessoni and Manoni 1996;

Thomas et al. 1997; Kaba et al. 2010a). This might be related to either low viral load in the blood or transient viremic status in the case of HEV, which might explain the higher impact of blood transfusion or organ transplantation as a risk factor for hepatitis E (Thomas et al. 1997; Khuroo et al. 2004; Boxall et al. 2006; Colson et al. 2007; Razonable 2011; Halac et al. 2012).

Finally, vertical transmission from mother to fetus has been reported frequently resulting in the death of the fetus (Khuroo et al. 2009; Aggarwal 2011).

3.4 Evidence of Zoonotic Transmission

Since the early 1990s, HEV antibodies have been detected in the sera of a variety of animals, such as monkeys, pigs, rodents, cattle, sheep, poultry, dogs, and cats, in both developed and developing countries (Wang et al. 2002; Chang et al. 2009; Meng 2010a; Wang and Ma 2010). Early after the discovery of HEV-like viruses in pigs in 1997, and a few years later in rodents (Favorov et al. 2000; Peralta et al. 2009b), the existence of endemic animal reservoirs was questioned with respect to the sporadic cases reported in humans in industrialized countries. Therefore, the possibility that the viruses causing human hepatitis E might be similar to animal strains and infect other species started being investigated by comparative molecular characterization of strains of different origin.

Presently, it is known that persons working in close contact with animals of species susceptible to HEV infection frequently result positive at testing for anti-HEV antibodies (Karetnyi et al. 1999; Meng et al. 2002; Chang et al. 2009; Masia et al. 2009). The ability of genotype 3 HEV strains to cross species barriers has been supported largely (Meng 2010a), and the foodborne transmission of animal viruses from pork, wild boar, and deer has been confirmed by a series of different evidence, as addressed in detail above.

Today, cross-species passage is thought to represent an important mode of transmission for zoonotic genotype 3 HEV, and may in fact be the main source for autochthonous HEV infection cases in North America and Europe (Meng 2010a; Pavio et al. 2010).

The ability of HEV to cross the species barrier has been confirmed by experimental infections of SPF (Specific Pathogen Free) piglets with a human genotype 3 virus, and by similar demonstration that swine genotypes 3 and 4 strains are able to infect non-human primates (Meng et al. 1998b; Arankalle et al. 2006; Ji et al. 2008), in support of the observations that genotype 3 strains typically detected in swine can naturally infect humans (Pavio et al. 2010). These viruses apparently cause mild or subclinical infection in primates, but also considering the high genetic variability within HEV, the possibility that specific strains within animal genotypes 3 and 4 are more virulent than others cannot be excluded. Since the molecular basis for host-pathogen interaction in HEV is still largely unknown, strains that normally present lower virulence might lead to a more severe course of disease in particular host conditions (Bouquet et al. 2012a).

Different levels of restriction in the cross-species infection with animal HEV strains exist, as was clearly shown in a recent study conducted on recently reported rat and rabbit HEV strains (Cossaboom et al. 2012). Using the pig model of experimental infection, injection with rat HEV did not result in any evidence of infection, as opposed to the rabbit HEV strain which induced consistent although low level viremia and fecal shedding of virus that was proven infectious when administered again to rabbits.

Finally, important evidence that supports the occurrence of zoonotic transmission of HEV has derived from the phylogenetic analysis of human and swine strains isolated in different regions of the world. Many studies have in fact reported full identity or close nucleotide and amino acid similarities between human and swine strains from the same geographic region (Banks et al. 2004a; Lu et al. 2006; Meng 2010b; Wenzel et al. 2011) that were often nearer than between strains which were detected from the same species but in different countries.

Chapter 4
Pathogenesis in Humans

In humans, hepatitis E virus (HEV) infection has been documented to use four possible routes of transmission: (1) waterborne, mostly in less developed areas; (2) foodborne, particularly through the consumption of raw or undercooked meat of infected domestic pigs or wild animals such as boar and deer; (3) blood-borne, or parenteral transmission; and (4) vertical transmission from mother to child.

Besides clinical observations during the course of natural disease in human patients, several elements of pathogenesis have been outlined on the basis of data from experimentally infected non-human primates. Under experimental conditions, infection can be transmitted by either the intravenous or oral routes; the latter, which in most cases is the natural infectious route, requires an infectious titer 10,000 times higher than the intravenous route (Emerson and Purcell 2003).

4.1 Incubation Period and Clinical Manifestations

The incubation period after oral exposure to HEV is about 4–5 weeks. The route and mechanism by which the virus reaches the liver from the intestinal tract remain unknown. Once in the liver, the virus replicates in the cytoplasm of hepatocytes, accumulates in the bile, and is subsequently shed in the feces. Viremia starts when the virus is already detectable in the liver (Kamar et al. 2012). However, it is not known to what extent extra hepatic sites of HEV replication occurs, and if all the virus present in the feces originates in the liver or may derive in part from replication in the intestinal tract. Viremia and fecal shedding are detected prior to liver abnormalities, which normally appear in concurrence with the humoral immune response and are characterized by an elevation of transaminases (Kamar et al. 2012). HEV can be detected in the stools beginning approximately 1 week before the onset of illness and can persist for as long as 2 weeks afterwards. Viral genomic RNA can be detected in the serum of most patients with acute hepatitis E for approximately 2 weeks, but in some infected subjects, its presence has been

F. M. Ruggeri et al., *Hepatitis E Virus*, SpringerBriefs in Food, Health, and Nutrition, 23
DOI: 10.1007/978-1-4614-7522-4_4, © Franco Maria Ruggeri, Ilaria Di Bartolo,
Marcello Trevisani, Fabio Ostanello 2013

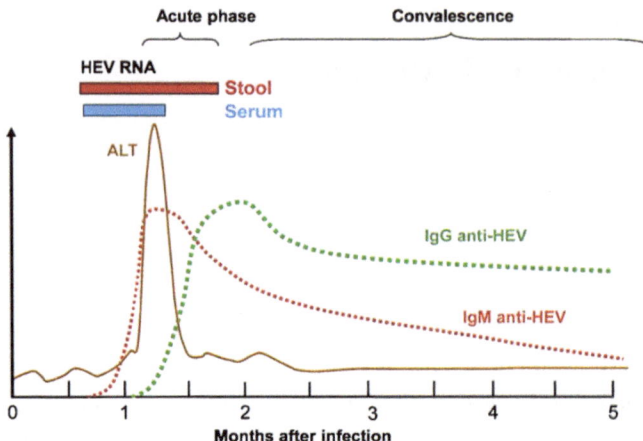

Fig. 4.1 Schematic summary of HEV infection, showing the main pathogenic events (virology, serology, disease) during the HEV infection. Reproduced from Krawczynski et al. (2011), with permission of Elsevier

reported for as long as 16 weeks after the onset of disease (Aggarwal and Jameel 2011) (Fig. 4.1).

The pathogenic mechanism leading to liver damage is not clear, but it is likely that immune-mediated elements are also relevant rather than simply relying on a direct action of the virus against the infected cell, which would be in line with the observation that HEV is not openly cytopathogenic. This hypothesis is supported by the fact that, during the course of the disease, infiltrating lymphocytes in the liver have been found to have mainly a cytotoxic/suppressor immune phenotype (Aggarwal 2011).

Anti-HEV IgM antibodies appear during the early phase of clinical illness, and disappear completely in 4–5 months. During an epidemic, IgM antibodies have been detected in more than 90 % of serum samples obtained from patients between 1 week and 2 months after the onset of illness. The IgG response appears shortly after the IgM response and the IgG antibody titer increases throughout the acute phase up to the convalescent phase, remaining high from 1 to 4 years after the acute phase of illness. The exact duration and persistence of anti-HEV antibodies is not known, but IgG antibodies have been detected up to 14 years post infection (pi) (Emerson and Purcell 2003; Aggarwal and Jameel 2011).

The clinical characteristics of hepatitis E can be very different. Acute icteric hepatitis, the most common recognizable form of illness, has an initial prodromal phase lasting a few days, with a variable combination of flu-like symptoms, fever, mild chills, anorexia, nausea, abdominal pain, vomiting, diarrhea, arthralgia, asthenia, and a transient macular skin rash. These symptoms are followed within a few days by darkening of the urine, lightening of the stool color and the appearance of jaundice. With the onset of jaundice, fever and other prodromal symptoms

Fig. 4.2 Histopathologic changes in the liver of cynomolgus macaques experimentally infected with HEV. Intralobular lesions showing focal necrosis, hepatocytic acidophilic degeneration, cell shrinkage, nuclear chromatin condensation, and multiple acidophilic body formation. Reproduced from Krawczynski et al. (2011), with permission of Elsevier

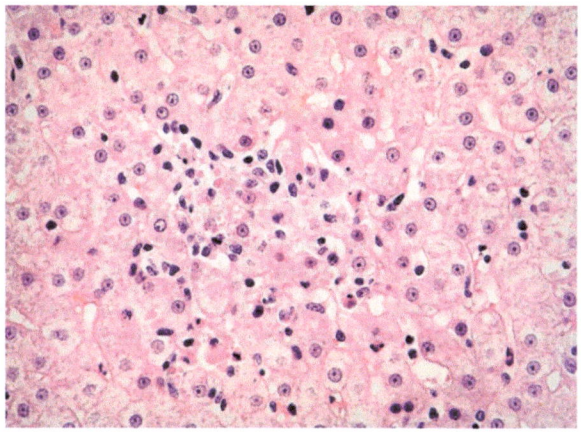

tend to diminish rapidly, and soon disappear entirely. Laboratory test abnormalities include bilirubinuria, variable degrees in the rise of serum bilirubin (predominantly conjugated), and a marked elevation in hepatic enzymes. The illness is usually self-limiting and typically lasts between 1 and 4 weeks; when it starts receding, the abnormalities of hematochemical parameters start receding, and gradually reach normal values (Aggarwal and Krawczynski 2000).

Hepatitis E is normally an acute disease, and until recently no evidence of chronic hepatitis or cirrhosis had been reported. However, a few patients were reported to have "prolonged" clinical illness with cholestasis, persistent jaundice, and prominent itching. In these cases, laboratory tests showed a rise in alkaline phosphatase and elevated bilirubin level persisting also after transaminase levels had returned to normal. The prognosis is normally favorable, and jaundice finally resolves spontaneously after 2–6 months (Aggarwal and Krawczynski 2000). Many infected individuals exhibit a mild clinical course, and develop only flu-like symptoms. In these patients, liver involvement is recognized only if laboratory studies are performed (Agrawal et al. 2012). However, a small proportion of patients have been reported to pass a more severe disease, with fulminant or subacute hepatic failure, sometimes culminating in the death of the subject (Fig. 4.2).

Recent studies have focused on a possible role of HEV as an important cause of chronic liver disease in immunosuppressed subjects, such as the recipients of organ transplants (Khuroo et al. 2009; Halac et al. 2012; Halleux et al. 2012; Koning et al. 2013), with possible involvement of the interferon transcriptome (Moal et al. 2013).

This completely different course of disease seems to match the higher clinical severity and exceptional mortality rate shown among pregnant women (Boccia et al. 2006; Patra et al. 2007; Aggarwal 2011). Pregnant women, particularly those in the second and third trimesters, are in fact more frequently affected during hepatitis E outbreaks, and the outcome is worse both for the mother and for the fetus or the newborn. Abortions, still births, and neonatal deaths have been reported.

Overall, the reason for the severity of disease in pregnant women remains unknown, and is probably not related to the poor health conditions frequently observed in developing countries, since the lethality rate in pregnant women in industrialized countries is also high.

The case-fatality rate in many reports has ranged from 0.07 to 2 %, which is slightly higher than that of hepatitis A. In its most benign form, HEV infection is entirely unapparent and asymptomatic, and would pass unnoticed. The actual frequency of asymptomatic infections is, however, not known, but it probably exceeds by far that of icteric disease. In endemic areas, a large proportion of individuals who were found to be positive for anti-HEV antibodies did not recall having had jaundice (Aggarwal and Krawczynski 2000; Aggarwal 2011).

4.2 Pathological Lesions

From the pathologist's perspective, the most common form of hepatitis E is a cholestatic-type of hepatitis that is characterized by canalicular bile stasis and by a gland-like transformation of parenchymal cells. In other patients, changes resemble those of other forms of acute hepatitis with regressive disseminated alterations of the hepatocytes (apoptosis, presence of ballooned hepatocytes and acidophilic cells, steatosis, focal hepatocyte necrosis), marked cholestasis with or without proliferation of bile ducts, Kupffer cell hypertrophy with accumulation of bile, inflammation of portal tracts with infiltration of neutrophils, macrophages, and lymphocytes (Aggarwal 2011).

Specific therapy against HEV is not yet available, although ribavirin has been used with some success in cases of acute liver failure, where a concomitant infection with HCV or other hepatitis viruses was not excluded (Goyal et al. 2012).

4.3 Immunity

The immune response elicited by HEV infections has not yet been fully elucidated. The virus is non-cytopathogenic; therefore, it is considered that liver injury may be the consequence of immune-mediated damage by cytotoxic T cells and natural killer cells. In fact, the possible involvement of cytotoxic T cells in disease pathogenesis during HEV infection was proven in the livers of patients with acute liver failure, where a significant infiltration of activated CD8(+) T cells containing granzymes was demonstrated (Prabhu et al. 2011). In patients with acute hepatitis E, altered numerical proportions of NK cells, NK cell subsets, and NKT cells were observed, but the alteration of NK and NKT cell numbers during the acute phase of the disease resulted reversible. Overall the data suggest that the innate immunity may play a relevant role in the pathogenesis of the infection (Srivastava et al. 2008).

In patients affected with hepatitis E, a serologic anti-HEV antibody response is usually detected at the time of onset of symptoms, with IgM appearing in the early phase of clinical illness and disappearing after 4–5 months. An IgG response then follows, with anti-HEV IgG titer increasing throughout the acute and convalescent phases, remaining elevated for years (Dawson et al. 1992; Favorov et al. 1992). Similar findings have been obtained in non-human primates infected experimentally.

Patients have also been shown to activate a PBMC-specific cellular immune response to HEV ORF2 protein, although not paralleled by an increase of HEV-reactive CD8 cells. In both humans with anti-HEV IgG seroconversion and in convalescent chimpanzees experimentally infected with HEV, the serum antibody titers and the cell mediated immune responses correlated well (Aggarwal et al. 2007; Shata et al. 2007; Srivastava et al. 2007).

A vaccine is not yet available in the Western world markets. However, since 2005 Chinese researchers have developed a recombinant vaccine from the original human Burmese strain of HEV, and tested in a randomized, controlled trial in over 110,000 individuals in a region of China at high risk of infection with genotypes 3 and 4 strains (Zhu et al. 2010). The trial has apparently shown the vaccine to have good tolerance and very high efficacy (99 %), and given its origin from a genotype 1 human strain it could also prove efficacious in industrialized countries where genotypes 1 and 3 HEV types circulate more widely.

Chapter 5
Epidemiology of HEV Infection in Animals

Significant progress has recently been made in understanding the natural history and animal reservoirs of HEV (Meng 2008; Pavio et al. 2010). Animal strains of HEV have been discovered in domestic pigs (Meng et al. 1997) and wild boar (Sonoda et al. 2004), chickens (Haqshenas et al. 2001), rabbits (Zhao et al. 2009), rats (Johne et al. 2010b), deer (Tei et al. 2003), mongooses (Nakamura et al. 2006), bats (Drexler et al. 2012), ferrets (Raj et al. 2012), and possibly also in cattle (Howard et al. 2010) and sheep (Wang and Ma 2010). This together with the existence of other animal species that are seropositive for HEV antibodies (Meng 2010a, b) have significantly broadened the host range and diversity of HEV. Recently, a novel member of the family *Hepeviridae* (called Cutthroat Trout Virus or CTV) was identified in cutthroat trout in the USA (Batts et al. 2011), and the virus presented a genome with marked similarities in both size and organization to that of typical hepatitis E virus, with which CTV shares higher nucleotide and amino acid sequence identities compared to any other virus family. However, the nucleotide sequence differences between CTV and hepeviruses from mammals or birds appeared to be sufficient to warrant the institution of a novel genus, within the family (Batts et al. 2011).

Screening for hepevirus RNA in more than 3,800 bat specimens from 85 different species and from five continents, Drexler and colleagues (Drexler et al. 2012) have found HEV-like viruses in African, Central American, and European bats. Based on genomic sequences, these viruses have been clustered in a novel phylogenetic clade within the family *Hepeviridae* (Drexler et al. 2012).

These findings of HEV-like viruses in the animal world is interesting and deserve further investigation, particularly because hepatitis E is now acknowledged as a zoonotic disease, with reservoirs in domestic pigs, wild boar, and likely other relevant animal species (Meng 2010a). Besides the obvious implications of the ubiquitous presence of HEV in animals relevant for the food production chain, such as domestic pigs and wild boar, its spread in several other animal species raises a public health concern for zoonotic infection also through direct contact

F. M. Ruggeri et al., *Hepatitis E Virus*, SpringerBriefs in Food, Health, and Nutrition, DOI: 10.1007/978-1-4614-7522-4_5, © Franco Maria Ruggeri, Ilaria Di Bartolo, Marcello Trevisani, Fabio Ostanello 2013

with infected animals, environmental contamination, particularly surface water, via animal HEV shedding in feces, as well as recreational and professional exposure hazards in the countryside (Meng 2010a, b; Pavio et al. 2010).

5.1 HEV Infection in Domestic Pigs

Swine HEV infections are spread both in industrialized and developing countries, where for many years the disease has been considered endemic in humans (Meng et al. 1997; Huang et al. 2004). Since the first identification of swine infection in 1997 in the USA (Meng et al. 1997), several other swine strains have been isolated in North and Central America, Asia, Europe, Africa, New Zealand, and Australia (Clayson et al. 1995; Chandler et al. 1999; Hsieh et al. 1999; Pina et al. 2000; Garkavenko et al. 2001; van der Poel et al. 2001; Yoo et al. 2001; Meng et al. 2002; Pei and Yoo 2002; Choi et al. 2003; Clemente-Casares et al. 2003; Takahashi et al. 2003; Banks et al. 2004a; Wang and Zhuang 2004; Caprioli et al. 2007b; Kaba et al. 2010c). At least two genotypes of swine HEV, genotypes 3 and 4, have been definitively identified and characterized from pigs worldwide and, as in the case of human strains, also swine strains have been shown to present a high degree of nucleotide and phylogenetic divergence from region to region. A short genotype 1 HEV-like sequence, otherwise considered to only infect humans, was reportedly detected in a pig from Cambodia (Caron et al. 2006), although independent confirmation of this report is lacking.

Swine strains, particularly those identified in industrialized countries, have often been related to human cases of disease in which no specific source of infection was identified (Haqshenas et al. 2001; Engle et al. 2002; Meng et al. 2002; Choi et al. 2003; Nishizawa et al. 2003; Takahashi et al. 2003; Banks et al. 2004c). In countries where HEV has been identified and serological studies have been performed, most pigs over the age of 3–4 months have been shown to possess HEV-antibodies (Clayson et al. 1995; Meng et al. 1997; Chandler et al. 1999; Hsieh et al. 1999; Pina et al. 2000; Garkavenko et al. 2001; van der Poel et al. 2001; Yoo et al. 2001; Meng et al. 2002; Pei and Yoo 2002; Choi et al. 2003; Clemente-Casares et al. 2003; Takahashi et al. 2003; Banks et al. 2004a; Wang and Zhuang 2004). The prevalence rate presents differences depending on the geographic region and the age of the animals. Swine younger than 2 months are usually seronegative or positive at low prevalence, whereas pigs over that age show seropositivity rates that often exceed 80 %. The approximate estimated prevalence rates in different types of sample in the swine (Emerson and Purcell 2003; Banks et al. 2004a) are reported in Table 5.1.

In recent years, virological surveys carried out in many countries (Leblanc et al. 2007; Di Bartolo et al. 2011; Di Bartolo et al. 2012) have detected HEV in high proportions (often >40 %) of samples from apparently healthy animals at slaughterhouses, near where they enter the pork production chain and are commercialized. The risk of HEV transmission associated with consumption of raw meat or due to contact with pigs has already been highlighted in other countries, as

Table 5.1 Approximate estimated prevalence rates in swine (S) and in wild boar (WB)

Country	Sampling year	Animals	Place	Sample	Prevalence (%)	Genotype	Seroprevalence (%)	Note	Refs.
Europe									
Austria	2007–2009	S	–		7.4	3		Samples from 81 pigs submitted for other routine diagnostic examinations	Zwettler et al. (2012)
Belgium	2008	S	Slaughterhouse	Feces	7	4	–	Pigs of 5–6 months	Hakze-van der Honing et al. (2011)
The Czech Republic	2010	S	Slaughterhouse	Feces / Liver / Meat	3 / 5 / 3	3	– / – /		Di Bartolo et al. (2012)
Denmark	2007–2008	S	Farm	Feces	49.5	3 or 4	73.2[1]	[1]40 farms (55 % positive); in sows	Breum et al. (2010)
France	2008–2009	S	Slaughterhouse	Sera / Liver	– / 4[2]	– / 3	31 / –	[2]Prevalence estimate taking into account the sampling design and sensitivity and specificity of the tests	Rose et al. (2011)
Germany	2007–2008	WB	Hunting	Liver	2.5	3			Kaba et al. (2010d)
	2010	S	Grocery stores	Liver	4	3			Wenzel et al. (2011)
	2007	WB	Hunting	Liver / Bile / Sera	38.1[4] / 56.3[4] / 15.7[4]	3	– / – / 29.9	Samples from 132 wild boar at different regional sites; [4]prevalence 68.2 % when considering HEV detection in at least one sample	Adlhoch et al. (2009)
Italy	2010	S	Slaughterhouse	Feces / Liver / Meat	41 / 6 / 6	3	– / – / –		Di Bartolo et al. (2012)
	2006	WB	Hunting	Bile	25	3	–		Martelli et al. (2008)

(continued)

Table 5.1 Continued

Country	Sampling year	Animals	Place	Sample	Prevalence (%)	Genotype	Seroprevalence (%)	Note	Refs.
Portugal	2010–2011	S	Farm	Feces	22	3	–	Five pig farms (all positive)	Berto et al. (2012c)
Serbia	2009	S	–	Liver	26	–	–	Samples from 50 pigs submitted for other routine diagnostic examinations	Savic et al. (2010)
Slovenia	2004–2005	S	Farm	Feces	20.3	3	–		
Spain	2010	S	Slaughterhouse	Feces	38	–	–	Six farms	Di Bartolo et al. (2012)
				Liver	3	–	–		
				Meat	0	–	–		
	–	S	Farm	Sera	18.8	3	20.4	85 pig farms (47 % positive)	Jimenez de Oya et al. (2011)
Sweden	2011	S	Farm	Feces	29.6	3	–	Piglets of 2–4 months; 22 farm	Widen et al. (2011)
		WB	Hunting	Sera	8.2	3	–		
Switzerland	2011	S	Slaughterhouse	Meat juice	–	–	49 – 60		Wacheck et al. (2012)
The Netherlands	2008	S	Slaughterhouse	Feces	15	3	–	Pigs of 5–6 months	Hakze-van der Honing et al. (2011)
	2005–2008	WB	Hunting	Sera	5	3	12		Ruijes et al. (2010)
				Liver	2		–		
				Feces	2		–		
				Muscle	0		–		
UK	2009–2010	S	Slaughterhouse	Liver	3	–	–		Berto et al. (2012a)
				Feces	13	–	–		
	2007	S	Farm	Feces	21.5	3	–	41 farms (100 % positive)	McCreary et al. (2008)
Hungary	2005–2009	S	Farm	Feces	21	3	–	41 farms (39 % positive)	Forgach et al. (2010)
				Liver	31		–		
		WB	Slaughterhouse	Liver	10.7		–		

(continued)

Table 5.1 Continued

Country	Sampling year	Animals	Place	Sample	Prevalence (%)	Genotype	Seroprevalence (%)	Note	Refs.
Asia									
Cambodia	–	S	Farm	Feces	21.3	3, 1	–	Pigs of 2–6 months	Caron et al. (2006)
				Sera	3.3		5.6		
China	–	S	Farm	Sera			78.9		Li et al. (2011)
			Slaughterhouse	Feces	8.3	3, 4			
				Liver	6.3	3, 4			
				Sera			79.6		
Korea	–	S	Farm	Feces	17.5	3		12 pig farms	Kim et al. (2008)
India	2007	S	Retail	Liver	0.8	4	–		Kulkarni and Arankalle (2008)
Indonesia	2008–2010	S	–	Sera	0.9[5]–1.2[6]	4	82.4[5]–61.8[6]	[5] Bali; [6] Java	Utsumi et al. (2011)
Laos	2009	S	Farm	Feces	11.6	4	–	Pigs ≤6 months	Conlan et al. (2011)
Japan	2003–2010	WB	Captured	Liver	3.4	3, 4 and unrecognized genotypes	–		Sato et al. (2011)
				Serum	2.2		8.1		
Mongolia	2006	S	Farm	Sera	36.6	3 types	91.8	Four farms, pigs of 2–3 months	Lorenzo et al. (2007)
Taiwan	1998–2000	S	Farm	Feces	6	3, 4[7]	–	Pigs of <2–>7 months[7] in pigs imported from USA	Wu et al. (2002)
				Sera	1.5		–		
Thailand		S	Farm	Sera	7.8	3	64.7	Five farms; pigs of 1–6 months and sows	Siripanyaphinyo et al. (2009)
Oceania									
Australia	–	S	Farm	Sera	–	–	30	Two commercial farms	Chandler et al. (1999)

(continued)

Table 5.1 Continued

Country	Sampling year	Animals	Place	Sample	Prevalence (%)	Genotype	Seroprevalence (%)	Note	Refs.
New Caledonia	2009	S	Farm	Feces	6.5[8]	3	–	Pigs of 1–6 months; one farm.[8] 18.8 % in 2–4 months pigs	Kaba et al. (2011)
New Zealand		S		Sera	–	–	75[9]	[9]22 farms (91 % herd seroprevalence)	Garkavenko et al. (2001)
				Feces	37.8[10]	3		[10] pigs of 1–3 months	
North America									
Canada	2003–2004	S	Farm	Feces	34[11]	3		70 farms in Quebec regions; pigs of 2–4 months; samples collected from the pen's floor.[11] Prevalence of positive farms	Ward et al. (2008)
			Slaughterhouse	Sera	–	–	59.4[12]	Pigs of 6 months;[12] mean prevalence; Quebec: 88.8 %; Ontario: 80.1 %; Alberta and Saskatchewan: 38.3 %	Yoo et al. (2001)
Mexico		S	Farm	Feces	31	3	–	Pigs of 2–4 months	Cooper et al. (2005)
				Serum	6		80		
USA	–	S	Farm	Sera	–	–	41.2		Dong et al. (2011)
South America									
Argentina	–	S	–	Sera	–	–	22.7	Swine older than 6 months	Munne et al. (2006)
			Farm	Feces	88.9	3	–	One farm, pigs of 1–2 months	
Bolivia	2006	S	–	Feces	31.8[13]	3	–	[13] of 22 swine fecal pools	Dell'Amico et al. (2011)

(continued)

Table 5.1 Continued

Country	Sampling year	Animals	Place	Sample	Prevalence (%)	Genotype	Seroprevalence (%)	Note	Refs.
Brazil	2008	S	Slaughterhouse	Bile	9.6	3	–		dos Santos et al. (2011)
	2009	S	Farm	Feces	15.3	3	–	14 farms (10 positive)	Gardinali et al. (2012)
Cuba	2013	S	Farm	Feces	18.8	3	–	Four farms	de la Caridad Montalvo Villalba et al. (2013)
Costa Rica	–	S	Farm	Feces	36.5	3	–		Kase et al. (2008)
Africa									
DR Congo	–	S	Farm	Fecal	2.5	3	–	Pigs tested are descendants of pigs imported from Belgium in 2002	Kaba et al. (2010b)
Madagascar	2010–2011	S	Slaughterhouse	Liver Sera	1.2	3	– 71.2	Pigs of 6 months	Temmam et al. (2013)

Note S swine; *WB* wild boar
Superscript numbers refer to special Notes

HEV RNA was detected in pig livers at grocery stores in Japan, the Netherlands, and the USA (Yazaki et al. 2003; Bouwknegt et al. 2007; Feagins et al. 2007a). In addition, HEV RNA or specific antibodies were detected in slaughterhouse workers or butchers in different countries (Dalton et al. 2007b; Galiana et al. 2008), suggesting that HEV transmission from pigs to humans during slaughter may also occur, and that contact with pigs may increase the risk of infection for workers (Galiana et al. 2010).

Once the possibility of zoonotic transmission is accepted, it appears clear that the higher the prevalence is in animals, the greater is the risk of transmission to humans. From this perspective, the prevalence data reported above represent serious reasons for concern about public health and the consumer.

5.2 HEV Infection in Wild Boars or Wild Pigs

To date, studies aimed at evaluating the presence of HEV RNA or specific antibody in wild boar or wild pigs have been conducted in Australia (Chandler et al. 1999), Japan (Matsuda et al. 2003; Yazaki et al. 2003; Sonoda et al. 2004; Takahashi et al. 2004; Tamada et al. 2004; Li et al. 2005b; Masuda et al. 2005; Nishizawa et al. 2005), France (Carpentier et al. 2012), Germany (Schielke et al. 2009), Spain (de Deus et al. 2008b; Boadella et al. 2012), Sweden (Widen et al. 2011), the Netherlands (Rutjes et al. 2010), and Italy (Martelli et al. 2008). However, the study in Australia was performed on wild hogs, whereas the Japanese investigations concerned the two wild boar subspecies *Sus scrofa leucomyxtas* and *Sus scrofa riukiuanus* (Watanobe et al. 1999) which are largely present in that country, but are phylogenetically different from the wild boar subspecies present in European countries. Sonoda and colleagues (Sonoda et al. 2004) reported that HEV RNA was present only in one pair of serum and liver specimens obtained from a sample of 41 animals, while Nishizawa and colleagues (Nishizawa et al. 2005) reported a prevalence of HEV-positive of 2.3 % out of a total of 87 animals. In Spain, de Deus and colleagues (de Deus et al. 2008b) reported that 27 out of 138 serum samples were RT-PCR positive for HEV, whereas in Italy 25 % of the 88 wild boar tested were reported to contain HEV RNA (Martelli et al. 2008). The different prevalence of HEV infection detected in this latter study (Martelli et al. 2008) may indicate a truly higher circulation of HEV in the investigated wild boar population, although differences due to the type of specimen and the PCR method chosen cannot be excluded. In other reports, detection of HEV was performed on either serum and/or liver samples, whereas Martelli and colleagues (Martelli et al. 2008) specifically examined bile samples, a type of sample that has also been recently reported by others to be the most reliable for the detection of HEV in pigs (de Deus et al. 2007). In addition, differences in the HEV prevalence among herds and countries might also be related to a different infectivity of the HEV strains specifically involved or to differences in the biology and ecology of the wild boar populations considered (genetics of the animals, density of the population, environmental characteristics). In this regard, it is of interest to note that the studies conducted on European wild boar by Widen et al. (2011) and by de Deus et al. (2008b) also revealed a prevalence of viremic animals on the order of 8 and 20 %, respectively. Concerning the possible risk factors associated with HEV infection, no significant differences in the virus prevalence have been detected when comparing differences in the management system (open: no fencing and no management; fenced: fencing and artificial feeding; intensive: livestock-like management) (de Deus et al. 2008b), sex and age classes (de Deus et al. 2008b; Martelli et al. 2008). Positive animals were detected in each age class, including juveniles of 4 months of age (Martelli et al. 2008), indicating that infection can occur at least starting from this age. The presence of HEV RNA in animals older than 24 months extends previous findings (Sonoda et al. 2004; Nishizawa et al. 2005) reporting HEV infection in wild boar of approximately 2 years of age. These data are in contrast with the results of most of the studies conducted on the domestic swine, which indicate that infection mainly occurs in animals of 3–5 months of age, has a short duration, and is generally self-limiting (Meng et al. 1997, 1998b). These differences might suggest that infection in wild boar can become chronic, possibly sustained by an incompletely protective immunity, or that there is continuous re-infection favored by a short-lasting immunity. However, the possibility that in the former studies the virus strain identified may have found a naïve population cannot be ruled out, with infection affecting all animals independent of age. Biometric characteristics (weight and length of the body) of the infected animals were not different than those of uninfected animals belonging to the same age and sex classes (Martelli et al. 2008).

Carpentier et al. (2012) reported that anti-HEV antibodies prevalence increased steadily from 10.4 % in animals aged less than 12 months to 15.7 % in older animals, but the difference was not statistically significant. In France, detection of anti-HEV antibodies varied with the geographical origin of the wild boar, being 7.3 % in the northern part of France, 9.0 % in the center, and 22.6 % in the southern part of the country (Carpentier et al. 2012). As HEV is excreted in the feces of infected animals, it can be speculated that HEV could be transmitted by contact with wild boar or deer (Rutjes et al. 2010) or their feces. As with other diseases of wild animals, the density of wild boar may have an important role in the epidemiology of HEV, and density variations could also explain the higher seroprevalence highlighted in some areas of France.

These results, together with the fact that the wild boar examined appeared to be clinically healthy, sustain the hypothesis that, as in domestic pigs (Meng et al. 1998b), HEV infection may commonly be subclinical in wild boar. These findings, together with the observation that HEV infection may be subclinical and can be present also in animals at an age in which they are commonly hunted to be eaten, are somewhat worrying because of the possible risk of transmission of HEV to human beings by either contact with infected boar or ingestion of contaminated undercooked meat or organs. In this respect, the presence of up to 25 % HEV-positive bile samples (Martelli et al. 2008) implies that at least wild boar liver can represent an organ at risk for zoonotic transmission. Nonetheless, it cannot be totally excluded that during the slaughtering process small amounts of HEV-positive bile might cross-contaminate other edible parts of the carcass.

Most strains of HEV recovered from wild boar worldwide belong to genotype 3 (Meng 2010a). However, recent studies conducted in Japan have reported the detection of novel HEV sequences belonging to genotype 4 (Sato et al. 2011) and to a new unrecognized HEV genotype (Takahashi et al. 2011) in wild boar.

5.3 HEV Infection in Other Wild Mammals (Deer and Mongoose)

Anti-HEV IgG antibodies were detected in approximately 2–3 % of samples taken from Sika deer (*Cervus nippon nippon*) and 35 % from Yezo deer (*Cervus nippon yesoensis*) in Japan (Sonoda et al. 2004; Matsuura et al. 2007; Tomiyama et al. 2009), and in 5 % of red deer (*Cervus elaphus*) in the Netherlands (Rutjes et al. 2010). The full-length genomic sequence of a strain of HEV was determined from a Sika deer in Japan, and sequence analysis revealed that the deer HEV belonged to genotype 3 (Takahashi et al. 2004). Genotype 3 strains of HEV were also genetically identified from roe deer (*Capreolus rufus* and *Capreolus capreolus*) in Hungary (Reuter et al. 2009; Forgach et al. 2010). Zoonotic transmission of HEV from deer to humans via the consumption of contaminated deer meat has been reported (Tei et al. 2003; Takahashi et al. 2004), and consequently the deer is to be considered an additional reservoir of HEV.

It has been reported that approximately 8–21 % of sera from mongooses examined in Japan were positive for anti-HEV antibodies, indicating relevant circulation of the virus in this wild animal species (Li et al. 2006; Nakamura et al. 2006). The full-length genomic sequence of a strain of HEV recovered from a mongoose was determined, and the virus was shown to be a genotype 3 HEV closely related to a genotype 3 HEV strain detected in a swine from Japan (Nakamura et al. 2006). The ability of the mongoose HEV to be transmitted across species and infect other animals or humans is unknown. However, the observation that mongoose HEV strains clearly belong to genotype 3 is highly suggestive that they may be zoonotic.

5.4 HEV Infection in Rats, Rabbits, Ferrets, and Bats

Rats, as well as other rodents, have been suspected for long time to represent a potential reservoir of HEV. Antibodies to HEV have been detected in various species of rats, including Norway (*Rattus norvegicus*), black (*Rattus rattus*), and cotton (*Sigmodon hispidus*) rats (Kabrane-Lazizi et al. 1999; Favorov et al. 2000; Arankalle et al. 2001; Hirano et al. 2003). In the 1990s, Kabrane-Lazizi et al. (1999) tested 239 wild rats trapped in different geographic regions of the USA for the presence of anti-HEV IgG antibodies, and found that approximately 44 % of the rats caught in Louisiana, 77 % in Maryland, and 90 % in Hawaii were seropositive. However, the virus responsible for anti-HEV seropositivity in rats could not be identified until recently (Johne et al. 2010b). By using a nested broad-spectrum RT-PCR, two authentic HEV sequences from fecal samples of wild Norway rats (*Rattus norvegicus*) from Hamburg, Germany (Johne et al. 2010b) have been amplified. Since then, rat HEV strains have been isolated from wild rats in other areas of Germany, and have been detected in wild rats also in the USA and Vietnam (Purcell et al. 2011; Li et al. 2013). The HEV strains characterized from rats shared only approximately 60 and 50 % of sequence identity with human HEV and avian HEV strains, respectively. Following their identification, the complete genomic sequences of two strains of rat HEV have been determined (Johne et al. 2010a), and sequence and phylogenetic analyses have revealed that the rat HEV belongs to a putative novel genotype within the genus Hepevirus of the family *Hepeviridae*. However, in a recent study conducted in the USA, sequences were obtained from HEV positive liver samples of wild Rattus spp., and all but one of these isolates was relegated to the zoonotic HEV genotype 3, whereas the recently discovered rat genotype from the USA and Germany was also detected (Lack et al. 2012). It remains to be determined if the rat HEV can effectively cross any species barrier and infect humans or other animals.

Recently, several new HEV strains have been isolated from farmed rabbits in China (Zhao et al. 2009; Geng et al. 2011c), and three full-length genomic sequences of the rabbit HEV were determined (Geng et al. 2011a). The rabbit HEV strains share approximately 85 % nucleotide sequence identity with each other and 74, 73, 78–79, 74–75, and 46–47 % nucleotide sequence identity to genotypes 1, 2, 3, 4 mammalian HEV, and avian HEV, respectively. Phylogenetic analysis revealed that the rabbit HEV is a distant member of the genotype 3 HEV. Approximately 57 % of the farm rabbits in Gansu Province, China were seropositive for anti-HEV antibodies with approximately 8 % of them also positive for HEV RNA. In a separate study from Beijing, China (Geng et al. 2010), approximately 55 % (65/119) of the farmed rex rabbits tested positive for anti-HEV antibody with approximately 7 % of fecal samples (8/115) also positive for HEV RNA. The rabbit HEV sequences recovered from rabbits in Beijing cluster together in genotype 3 with those from Gansu Province. Since the rabbit HEV belongs to genotype 3, it is thus possible that the rabbit HEV may be zoonotic. Additional studies are warranted to determine the host range and species tropism of the rabbit HEV.

In the Netherlands, 9.3 % of fecal samples collected from household pet ferrets that did not show overt clinical signs were found positive for HEV RNA (Raj et al. 2012). Phylogenetic analysis of the complete genomes clearly showed that HEV from ferret was separated from genotypes 1–4 HEV.

As mentioned above, novel hepeviruses have been described by comparative analysis of genome sequences from a globally representative sample of 3,869 bat specimens (Drexler et al. 2012). The genomic characterization of bat hepeviruses clearly supported their classification as members of the family *Hepeviridae*, and indicated that they may be the most divergent mammalian hepeviruses described so far. Bat hepeviruses appeared to be highly diversified, with a sequence variation comparable to that exhibited by human HEV strains. No evidence for the transmission of bat hepeviruses to humans has been found by analysis of over 90,000 pooled human sera from blood donors and individual patient sera collected in Cameroon and Germany in 1998 and 2009–2010, respectively. Full-genome analysis confirmed formal classification of bat hepevirus within the family *Hepeviridae*. However, sequence analysis and distance-based taxonomic evaluations suggest that bat hepeviruses constitute a distinct genus within the family *Hepeviridae*. Altogether, these findings suggest that hepeviruses may have first appeared in mammalian hosts a long time ago, and have subsequently undergone differentiation into genera according to different host restrictions. Human HEV-related viruses found in farmed and peridomestic animals might thus represent more recent secondary acquisition of human viruses by these animal species, rather than being animal precursors causally involved in the evolution of human HEV (Drexler et al. 2012).

5.5 HEV Infection in Avian Species

Avian HEV was first identified in chickens with hepatitis-splenomegaly syndrome (HSS) in the USA (Haqshenas et al. 2001), and later it was reported in chickens with the same disease in Canada (Agunos et al. 2006). In Europe, outbreaks of HSS have also been reported (Morrow et al. 2008), and particularly a widespread infection of avian HEV was described in chickens in Spain (Peralta et al. 2009a).

The disease is characterized by an increase in the rate of mortality among broilers and laying chickens, and was found to be responsible for decreases in egg production of up to 20 %. Regressive ovaries, red fluid in the abdomen, and enlarged liver and spleen with histological changes of hepatic necrosis and hemorrhaging were often reported in infected chickens (Sun et al. 2004). In experimental conditions, the avian HEV isolated from chickens was shown to be also infectious for turkeys.

Unlike swine HEV, which have been demonstrated to infect non-human primates in experimental conditions (Meng et al. 1998a, b), avian HEV could not be transmitted to monkeys (Huang et al. 2004; Sun et al. 2004), thus suggesting that, differently from swine and other animal strains, avian HEV most likely is not transmissible to humans. In fact, HEV avian strains studied to date have been reported to share about 50–60 % nucleotide sequence identity with human and swine HEV strains (Huang et al. 2004). Conversely, an approximately 80 % nucleotide identity was found between avian HEV and the big liver and spleen disease (BLSV) virus, previously identified in Australian chickens (Payne et al. 1999; Haqshenas et al. 2001). The avian HEV and BLSV strains might thus represent variants of the same virus (Haqshenas et al. 2002; Huang et al. 2004; Guo et al. 2006).

The genome of avian hepeviruses shows a similar organization to other animal HEVs, but its shorter length (6.6 kb) and low sequence identity with other hepeviruses of humans and animals (approximately only 50 %, at the nucleotide level) appear sufficiently divergent to support the existence of a separate genus (Marek et al. 2010). The latter is considered to contain three distinct genotypes, but a putative novel genotype has been recently proposed in Hungary, suggesting that the avian HEV diversity may in fact be higher than previously thought (Banyai et al. 2012).

Although avian HEV is unlikely to be a zoonotic virus, it is nonetheless an important problem for avian species farming, and for this aspect too, it is proving very different from the zoonotic swine HEV infection in the pig. In fact, whereas the latter infects pigs in the complete absence of symptoms (Meng et al. 1998a, b), avian HEV appears to induce a markedly symptomatic disease in chickens (Billam et al. 2005; Agunos et al. 2006) that may culminate in relevant production and economic losses in flocks where HSS spreads (Ritchie and Riddell 1991).

5.6   Other Potential Animal Reservoirs for HEV

Antibodies to HEV have been detected from 4.4 to 6.9 % of cattle in India (Arankalle et al. 2001), and 6–93 % of cattle in China (Wang et al. 2002; Zhang et al. 2008; Yu et al. 2009; Geng et al. 2010). In a prospective study of six newborn calves looking for evidence of HEV infection in the USA (Goens et al. 2003), it was shown that the first calf began to seroconvert for anti-HEV IgG antibodies at about 3 months of age, and that by the age of 7 months, a clear anti-HEV seroconversion was present in five out of the six animals. However, any attempts to identify HEV specific genome sequences in these cattle were unsuccessful (Goens et al. 2003), not excluding the presence of cross-reactive antibodies by molecular mimicry with proteins of a different nature. More recently, however, a 189 bp ORF2 sequence of HEV was reportedly amplified from fecal samples collected from eight cows in two distinct farms in China (Howard et al. 2010). The eight HEV sequences from Chinese bovines shared 96–100 % nucleotide sequence identity with each other, and 76–86, 82–84, 79–85, 84–96 % nucleotide sequence identity with the mammalian HEV strains of genotypes 1, 2, 3, and 4, respectively. Based on the short sequence available, it appears that the bovine HEV may belong to genotype 4 of HEV (Howard et al. 2010). However, thus far this is the only report on bovine hepevirus, and the authenticity of these genotype 4-like HEV sequences from cattle in China is still lacking an independent confirmation. Few studies reported detection of HEV RNA in sera from horses, data indicated that horses can acquire HEV infection and suggest that horses can represent a further reservoir of infection (Saad et al. 2007; Zhang et al. 2008). In Japan, a long-term monitoring of HEV infection in monkeys housed in a facility showed an outbreak of HEV among animals and detection of HEV RNA that showed it was related to genotype 3 strains circulating in the same country (Yamamoto et al. 2012). Serological evidence of HEV infection in sheep has also been reported.

Approximately 10–12 % of the sheep tested in China (Zhang et al. 2008; Chang et al. 2009) and 2 % of sheep tested in Spain (Peralta et al. 2009b) were reported to be positive for serum anti-HEV IgG antibodies. In addition, a short 189 bp HEV-like sequence was amplified from fecal samples from six sheep in China (Wang and Ma 2010). These six short HEV sequences from sheep were 99–100 % identical to each other, but shared only 79–85, 81–83, 79–84, and 85–95 % nucleotide sequence identity to the genotypes 1, 2, 3, and 4 of mammalian HEV strains, respectively. Based on the short sequence available, the sheep HEV appears to be a genotype 4 HEV (Wang and Ma 2010). However, as in the previous case of HEV sequence identification in the bovine, neither these short genotype 4-like HEV sequences in sheep have been independently confirmed by other groups.

In addition to the animal species from which HEV strains have been identified by molecular detection methods, as presented above, serological evidence of HEV infection has also been reported in a number of other animal species including dogs (Arankalle et al. 2001; Zhang et al. 2008; Liu et al. 2009), cats (Kuno et al. 2003; Song et al. 2010), goats (Shukla et al. 2007; Peralta et al. 2009b; Geng et al. 2010), horses (Zhang et al. 2008), and rhesus monkeys (Tsarev et al. 1995). Other studies failed to detect anti-HEV serum antibodies in some species, such as goats (Arankalle et al. 2001; Wang et al. 2002). Thus far, the real basis of anti-HEV seropositivity in these animal species has not been identified with certainty. In fact, virus or HEV-specific genome sequences were not recovered or were only detected sporadically in samples from these animal species. Thus, it is likely that more and new animal strains of HEV exist, but more studies are needed before the natural history of HEV may be considered fully disclosed.

5.7   Routes of Transmission and Risk Factors in Swine

A comprehensive picture of all natural methods of transmission of HEV between animals and across and between species, is not yet available, including the pig where most of the information on HEV comes from. Swine HEV can be transmitted experimentally via direct contact with infected pigs (Meng et al. 1998a; Kasorndorkbua et al. 2004). Transmission of HEV infection from infected to uninfected animals kept in direct contact has been clearly demonstrated, thereby confirming the contagiousness of the virus (Meng et al. 1998a, b; Kasorndorkbua et al. 2004). Fecal-oral transmission may also occur, but is difficult to reproduce in experimental conditions and may require repeated virus exposure. For this reason, the intravenous route is usually preferred in experimental transmission studies (Meng et al. 1998a).

Daily repeated direct contact among pigs reared in confinement buildings may enhance the spread of swine HEV. Pigs housed in the same pen are exposed to saliva, nasal secretions, urine, and feces of multiple pen mates, an event that is performed repeatedly each day.

Positive-strand HEV RNA was detected from 3 to 27 days post intravenous inoculation in various tissues including livers, lymph nodes, colons, small intestines, stomachs, spleens, kidneys, tonsils, salivary glands, and lungs from pigs inoculated with swine HEV (Williams et al. 2001). This finding is, therefore, suggestive of the existence of extra-hepatic sites of HEV replication in the swine, including particularly the intestinal tract. Taking all these data together, it is a reasonable assumption that under natural conditions swine HEV is transmitted via the fecal-oral route, as is thought to be the case in human HEV infections (Emerson and Purcell 2003). However, other route(s) of transmission cannot be ruled out.

Repeated use of needles for drug administration or vaccination is commonly practiced in swine health management. Even though HEV viremia in swine is transient (1–2 weeks), it is possible that blood contamination of needles from viremic pigs may also play a role in the spreading of HEV among pigs on the same and between farms (Kasorndorkbua et al. 2004). However, although this and other possible mechanisms were proposed (Kasorndorkbua et al. 2004), a conclusive demonstration of HEV transmission by either contaminated needle exposure or tonsil and nasal secretions could not be achieved.

HEV presence was detected also in urine specimens in contact-infected (Bouwknegt et al. 2009) and naturally infected pigs (Banks et al. 2004c). A HEV RNA-positive urine sample was obtained from a contact-infected pig at 65 days after onset of fecal excretion. This finding may suggest that HEV excretion via urine lasts longer than fecal HEV excretion, and/or that urinal HEV excretion occurs at a later stage of infection (Bouwknegt et al. 2009). However, no experimental study was conducted to demonstrate that the virus transmission can occur by exposure to urine, and the HEV RNA detected may represent virus that is bound to antibodies for disposal; anti-HEV antibodies have in fact been observed in urine samples of human patients with hepatitis E (Joshi et al. 2002).

In order to assess if HEV presents health and reproductive issues in pigs similar to those identified in humans, pregnant gilts were intravenously inoculated with swine HEV (Kasorndorkbua et al. 2003). Hepatitis E virus RNA was detected in the feces of all gilts and also at various times in the serum and liver, but clinical signs of disease were not evident. In addition, in two of 14 farms tested in South Korea HEV RNA has been detected simultaneously in the livers of aborted fetuses and in fecal and serum samples from their sows. HEV RNA was respectively present in 12.5–26 % of livers from aborted fetuses, in 15.4–50 % of sera, and in 16.6–25 % of fecal samples from sows (Hosmillo et al. 2010).

All of the HEV-positive fetuses were positive for Porcine circovirus type 2 (PCV2) but negative for other abortigenic viruses. These results indicated the occurrence of transplacental HEV infection in field pig farms. Although PCV2 alone is sufficient to cause reproductive failure, other case reports and retrospective analyses have implicated PCV2 in conjunction with other reproductive disease agents (Pensaert et al. 2004). Moreover, PCV2-affected pigs can become immunosuppressed, which facilitates secondary infections. Therefore, it is conceivable that transplacental HEV infection can be triggered by a concurrent PCV2 infection. This would explain the previous inability to reproduce HEV infection in pregnant pigs.

The risk factors at either the individual or farm level, and the transmission dynamics of HEV infection thus require further investigation.

Field cross-sectional studies conducted on different countries (Fernandez-Barredo et al. 2007; Di Bartolo et al. 2008) have shown that the prevalence of infected farms in a country can be very high, varying between 10 and 100 %. However, epidemiological comparison is very difficult due to the different sampling methods, expected prevalence, diagnostic procedures used, and the different farms classification. Also, epidemiological information on intra-herd prevalence may be influenced by the type of animals examined (breeding, fattening pigs), the age of the pigs, and the characteristics (farrow-to-weaning, farrow-to-finish) and size of the farm.

Di Bartolo and colleagues (Di Bartolo et al. 2008) have reported that HEV prevalence ranged from 12.8 to 72.5 % between farms in Italy, independent of farm type, with a higher prevalence in herds having >1,000 sows. This latter finding is common for most swine infectious diseases and, in addition to the number of susceptible animals, may also be related to the density and proximity between individuals (Nakai et al. 2006), and to the increased HEV shedding by pregnant sows (Fernandez-Barredo et al. 2006). Other variables such as the frequency of introduction of animals, the number of suppliers, could also influence virus spread inside bigger farms. HEV prevalence in Italy was found to be higher in fattening than in farrow-to-weaning and farrow-to-finish farms (Di Bartolo et al. 2008; Martelli et al. 2010). Within the herd, virus shedding with feces was observed in pigs of all age categories, although prevalence of HEV in fatteners was lower than in the younger weaners (Fernandez-Barredo et al. 2007; Di Bartolo et al. 2008).

Previous studies on serum and fecal samples have demonstrated that HEV RNA can be primarily detected in pigs of 2–5 months of age, whereas animals younger than 2 months were generally negative (Meng et al. 1997, 1998b) or had low prevalence. Fernandez-Barredo and colleagues (Fernandez-Barredo et al. 2006) reported that in farrow-to-weaning, grower-to-finish, and farrow-to-finish farms, only 11 % of the 3-week-old animals were proven to shed the virus. The prevalence increased in weaners (42 %), reaching a maximum (60 %) at 13–16 weeks of age. This, together with the knowledge that maternal immunity lasts approximately 2 months, makes occurrence of natural infection more likely to occur at approximately 1–2 months of age (Nakai et al. 2006; Satou and Nishiura 2007).

Viremia has been reported to last for 1–2 weeks after infection, and virus excretion for about 3–4 weeks, followed by seroconversion (Meng et al. 1998a; Nakai et al. 2006). Therefore, animals older than 6–8 months are expected to have cleared HEV infection. However, in several studies HEV has been detected in feces of animals of all age categories, from suckling piglets to old sows, well beyond 6–8 months of age (Fernandez-Barredo et al. 2006; Di Bartolo et al. 2008). A fairly high HEV prevalence has also been reported among breeders in Spain (Fernandez-Barredo et al. 2006, 2007), and to a lesser extent, in 5–6 month old swine in Japan (Nakai et al. 2006). In particular, Leblanc and colleagues (Leblanc et al. 2007) found fecal HEV in 41.2 % of slaughtered swine, at an age of 22–29 weeks, suggesting that the swine may possibly retain its susceptibility to HEV infection at any age. These findings are consistent with a report from the USA (Feagins et al. 2007a) showing that 11 % of commercial pig livers from the grocery store contained HEV RNA of genotype 3. Infection in pigs could be more prolonged than was previously thought or might become chronic, possibly sustained by an incomplete protective immunity. However, particularly in a virus loaded environment (Fernandez-Barredo et al. 2006), the possibility of continuous re-infection favored by a non-protective or short-lasting immunity may not be excluded. The latter hypothesis may fit with the finding that the high HEV prevalence among weaners and gilts seemed to decrease in fatteners and young sows, but was higher in old sows than in other age groups (Di Bartolo et al. 2008). These findings could otherwise be explained by a recent introduction of HEV into previously naïve herds. However, both the simultaneous occurrence on distinct farms and the identification of a variety of HEV strains make such an event very unlikely (Di Bartolo et al. 2008). Rather, it is more reasonable to assume that new different HEV strains or variants can be continuously introduced onto farms where pigs may be incompletely protected against novel HEV strains. Although newly introduced animals may be the most probable vehicles for novel viral strains onto a farm, contaminated feed, the environment or humans might also play a role, in particular in farrow-to-finish or farrow-to-weaning farms. Biosecurity measures and control of animal import may thus be important to prevent new virus introduction.

5.7.1 Transmission Dynamics of HEV in Pigs

The transmissibility of an infection can be quantified by its basic reproductive number R_0, defined as the mean number of secondary infections seeded by an infective animal into a completely susceptible (naïve) host population. For many simple epidemic processes, this parameter determines a threshold: whenever $R_0 > 1$, an infectious individual typically gives rise, on average, to more than one secondary infection, thus leading to an epidemic. In contrast, when $R_0 < 1$, each infectious individual typically gives rise, on average, to less than one secondary infection and the prevalence of infection cannot, therefore, increase (Cintron-Arias et al. 2009). The circulation of the virus will eventually decline. Field studies (Fernandez-Barredo et al. 2006; de Deus et al. 2008b; Di Bartolo et al. 2008; Casas et al. 2011) have shown a peak prevalence of HEV RNA in pigs of intermediate age, and a non-zero prevalence in finishing pigs at slaughter age. This prevalence pattern can be understood using a model that describes the transmission between pigs; the model can also be used for simulating the effects of vaccination on the fraction of infectious animals at slaughter.

The study by Satou and colleagues (Satou and Nishiura 2007) was the first report on HEV transmission estimated from field data: studying HEV transmission in six different Japanese provinces, these authors found the reproductive number to be in the order of 4.02–5.17, which is in agreement with the estimated reproductive numbers determined by Berto et al (2012a). However, the R_0 values calculated by Berto and colleagues (Berto et al. 2012a) were lower than the R_0

values previously calculated by Bouwknegt and colleagues in experimental infection conditions (Bouwknegt et al. 2008). This inconsistency is most likely related to the different transmission rate parameters between the experimental and field situations, even though the infectious periods in the two investigations were comparable.

In the very first HEV transmission dynamics study in an animal experiment, the R_0 was found to be either 8.8 or 32 in two separate experiments (Bouwknegt et al. 2008). The difference observed can likely be explained by the closer proximity of animals in an experimental setting compared to a farm situation, and by the fact that contact animals in a transmission experiment encounter only animals that are in the early, and possibly more infectious, stages of virus shedding that can also vary between distinct experiments.

In conclusion, altogether these studies reported that the estimated R_0 for HEV in swine is consistently higher than 1, indicating that HEV is certainly able to spread among pigs. As a consequence, pigs can be considered to have characteristics of infection making them an actual animal reservoir.

Sufficient information on the possible strategies to be applied for control of HEV infection on swine farms is not available presently. However, given the analogy with human beings, where HEV antibodies seem to be protective (Wibawa et al. 2004), the possible use of indirect control strategies to prevent the virus from spreading within herds, and eventually to minimize risks for human health should not be overlooked in the future.

Backer et al. (2012) evaluated the fraction of HEV RNA positivity at slaughter age in swine, and explored the effects of vaccination in this regards. When vaccination reduced the mean infectious period, the infectious fraction decreased.

5.7.2 Disease Associated with HEV Infection

The possible interaction between HEV and other pig pathogens has also to be fully clarified. In particular, it has been suggested that HEV infection could act in synergy with Porcine circovirus type 2 (PCV2), a virus also able to cause hepatitis in pigs (Segales et al. 2005). The original description of liver lesions in pigs infected with the prevalent hepatitis E virus strains were very similar to those typical of hepatic disease associated with PCV2, including the presence of syncytial cells in lymphoid tissue (Meng et al. 1997). Whether or not PCV2 was or was not a cofactor in the development of these lesions has not been clarified (Ellis et al. 2004).

Martelli and colleagues (Martelli et al. 2010) reported that no significant association could be observed to correlate the presence of HEV RNA in pigs with the concomitant PCV2 and/or PRRSV infections detected at the time of death, although the HEV prevalence seemed to be slightly higher in PCV2/PRRSV positive swine. These data are consistent with a previous study (de Deus et al. 2007) that reported no statistically significant difference in the HEV prevalence between PCV2-infected and non-infected pigs. Likewise, no significant associations were detected between the presence of HEV infection and the occurrence of any spe-cific pathological lesion (liver, intestine, lung, or other disorders), as detected in the animals at necropsy, not even in the case of lesions affecting the liver (Martelli et al. 2010). These results confirmed the apparent asymptomatic nature of HEV infection in pigs, at least at a macroscopic level (Meng et al. 1998a; Kasorndorkbua et al. 2002; Leblanc et al. 2007). The significant increase of hepatic histopathological lesions constantly associated with HEV infection (Meng et al. 1998a; Halbur et al. 2001; Williams et al. 2001; Lee et al. 2007; Martin et al. 2007) should be investigated in more detail, and possible less evident effects and damage caused by this apparently subclinical infection on the pig productivity and performance (e.g. growth rate, feed intake, nutrients digestion, pig carcass quality, reproductive parameters) should be further studied, especially in intensive rearing system farms.

Finally, additional studies should desirably be undertaken to evaluate HEV pathogenesis in more detail, in particular when HEV infection is associated with infections with other viral agents, such as Porcine circovirus type 2 (PCV2) and porcine reproductive and respiratory syndrome virus (PRRSV) (Cho and Dee 2006; Martin et al. 2007). These may help understand the possible role played by HEV in the genesis and evolution of multifactorial or conditioned diseases.

Chapter 6
Pathogenesis in Swine

6.1 Incubation Period and Clinical Manifestations

The pathogenesis of swine HEV is largely unknown. As is the case for humans, in the pig it is also still not clear how the virus, once it has entered the host, can reach the liver and which are its primary replication sites. Although hepatocytes are the primary site for HEV detection (Meng et al. 1997, 1998b; Williams et al. 2001), little is known about viral detection and localization in the liver and extra-hepatic tissues (Choi and Chae 2003; Ellis et al. 2004; de Deus et al. 2007). Using polymerase chain reaction (PCR) or in situ hybridization in animals experimentally infected by the intravenous route, it is possible to detect viral RNA in several extra-hepatic tissues such as lymph nodes, tonsils, spleens, stomachs, kidneys, salivary glands, lungs, and small and large intestines up to 20–27 days pi, also in the absence of viremia (Williams et al. 2001). However, the negative strand form of the virus genome, which is its replicative form, has not only been detected in the liver but principally in the intestine and in the lymph nodes (Williams et al. 2001), between 7 and 27 days pi. Viremia can last for about 2 weeks, but the virus can be detected in the feces for a longer time, from 3 to 50 days pi (Lee et al. 2009a). Seroconversion occurs about 2–3 weeks pi (Meng et al. 1998a, b; Halbur et al. 2001; Williams et al. 2001). Tissues in which the virus first replicates and persists (from 3 to 27 days pi) are the liver, the small intestine, the colon, and the lymph nodes (Halbur et al. 2001; Williams et al. 2001). These observations, together with the fact that during experimental infections viral RNA can be detected in feces earlier than in the bile, and in quantities that are tenfold higher, have led to the speculation that, after entering by the oral route and before inducing viremia, the virus replicates in the gastrointestinal tract (Meng et al. 1998a, b). Virological studies conducted on samples of serum and feces from swine herds have shown that HEV RNA can be primarily detected in pigs of 2–5 months of age, whereas animals younger than 2 months and older than 6–8 months are generally negative (Hsieh et al. 1999; Pina et al. 2000; van der Poel et al. 2001; Yoo et al. 2001; Meng et al. 2002; Pei and Yoo 2002; Choi et al. 2003; Clemente-Casares et al.

F. M. Ruggeri et al., *Hepatitis E Virus*, SpringerBriefs in Food, Health, and Nutrition,   49
DOI: 10.1007/978-1-4614-7522-4_6, © Franco Maria Ruggeri, Ilaria Di Bartolo,
Marcello Trevisani, Fabio Ostanello 2013

2003; Takahashi et al. 2003; Banks et al. 2004a). Considering these observations and given that maternal immunity is believed to last about 2 months, natural infection is thought to occur as early as approximately 1–3 months of age (Halbur et al. 2001; Kasorndorkbua et al. 2002; Meng et al. 2002).

Viremia follows initial infection, lasting for 1–2 weeks, and the virus is then excreted in the feces for about 3–4 weeks (at 3–5 months of age) with subsequent seroconversion and clearance of the infection by the immune system (Meng et al. 1998a, b; Hsieh et al. 1999; Haqshenas et al. 2001). HEV infection would, therefore, be relatively short, and would be self-limiting in only a few weeks (Meng et al. 2002; Yazaki et al. 2003).

Pigs of any age are asymptomatic, whether infected with swine HEV naturally or experimentally (Meng et al. 1998a; Kasorndorkbua et al. 2002; Leblanc et al. 2007; dos Santos et al. 2009).

6.2 Pathological Lesions

The hepatitis E virus, at least the known genotype 3 strains, appears to be attenuated or at least not particularly virulent for domestic swine, while there is no information available on its virulence in wild boar. The study of the disease in naturally or experimentally infected pigs has shown that HEV normally leads to subclinical infections with signs of hepatitis which can be detected only at histological analysis (Meng et al. 1998a, b, 2002; Halbur et al. 2001; Williams et al. 2001). Some studies have highlighted a possible correlation of the infection with liver damage (Lee et al. 2007; dos Santos et al. 2009) and histopathological lesions have often been detected in the liver of experimentally and naturally infected animals (Meng et al. 1998a; Halbur et al. 2001; Lee et al. 2007, 2009a; Martin et al. 2007; de Deus et al. 2008a). These lesions are characterized by mild multifocal sinusoidal and periportal lymphoplasmocitary infiltrations and small areas of vacuolar degeneration and necrosis of hepatic cells (Meng et al. 1998a, b, 2002; Halbur et al. 2001; Williams et al. 2001). Hepatic inflammation and hepatocellular necrosis peaked in severity at 20 days pi (Halbur et al. 2001). In addition, slight subepithelial lympho-histiocytic cell infiltrations were observed in the gall bladder at 28 days pi (Bouwknegt et al. 2009). An increasing trend in AST and ALT enzyme levels was also observed, which might suggest a slowly progressive development of liver damage occurring during HEV-infection (Bouwknegt et al. 2009).

In some cases, lymphoplasmocitary enteritis and multifocal interstitial lymphoplasmocitary nephritis have been detected. In experimentally infected animals, macroscopic lesions, such as mild to moderate enlargement of the mesenteric and hepatic lymph nodes, have sometimes been detected (Halbur et al. 2001). In the ileum a mild or moderate hyperplasia of Peyer's patches was observed. No lesions were observed in the spleen or pancreas (Bouwknegt et al. 2009).

It is of interest to note that human strains experimentally inoculated in pigs usually led to more severe histopathological lesions than swine strains (Meng et al. 1998a; Halbur et al. 2001). This is suggestive that some genotypes or strains of HEV might exhibit very different virulence in one or more hosts.

Some authors have suggested that the virus in pigs could behave in the same way as the human hepatitis A virus, which is virulent only in adults and not in juveniles (Meng et al. 1998b, 2002). This means that the disease would not be able to develop clinical symptoms because the majority of the adult pigs would be protected from infection, having encountered the virus at a juvenile stage and consequently having developed protective active immunity (Meng et al. 2002).

In a recent study, 12 seronegative gilts were inoculated intravenously with a swine HEV strain, and no clinical signs or fever were observed in the inoculated gilts or their fetuses during the whole experimental period. Mild multifocal lymphohistiocytic hepatitis was observed in four of 12 inoculated gilts, while there was no significant effect of swine HEV on fetus size, viability, birth weight, or weight gain in the offspring (Kasorndorkbua et al. 2003).

It is thought that HEV, although initially appearing as a mild pathogen, may nonetheless be able to lead to a severe disease when acting in synergy with other viral agents, such as Porcine circovirus type 2 (PCV2) (Ellis et al. 2004).

6.3 Immunity

From a serological point of view, passive maternal immunity has been shown to decline after 3–4 weeks of age and to terminate after 8–9 weeks; piglet IgG seropositivity was associated with the serological status of the sows (de Deus et al. 2008a). As previously mentioned, it is uncommon to find seropositive animals below the age of 1–2 months, however, that could have been biased by the low sensitivity of the serological tests used when some of these studies were performed (Meng et al. 2002).

The average time until antibody development was found to be 2 weeks after the first HEV RNA excretion in feces.

After infection, pigs develop an immune response against the virus, which is characterized by an early IgM antibody response followed by an increase of the IgG, after about a week. IgM antibody titer decreases rapidly (in 1–2 weeks), while IgG concentration raises constantly for several weeks.

The time determined before IgG development for intravenously inoculated pigs was 2 weeks post-inoculation at the earliest (Meng et al. 1998a; Halbur et al. 2001), but more frequently times between 3 and 8 weeks post-inoculation are reported (Meng et al. 1998a).

In the field, seroconversion can be detected after the viremic phase at approximately 3–4 months of age (antibodies peak at 4 months), and animals show high serological titers until 5–6 months of age. At that time, the IgG titer starts to decline slowly (Hsieh et al. 1999; Haqshenas et al. 2001; Williams et al. 2001; Yoo et al. 2001; Meng et al. 2002).

Although no strategic plan for vaccination of pigs against HEV infection has been discussed by veterinary health policy makers, immune prophylactic control might turn out to become desirable if zoonotic hepatitis E spreads more in the future. Pilot experiments have addressed these aspects, indicating that a vaccine approach may ensure an efficient cross-reactive protection in swine (Sanford et al. 2011).

Chapter 7
Diagnosis of Hepatitis E Infection

Diagnosis of hepatitis E should be considered an option in any patient who presents with a clinical evidence of hepatitis. The symptoms in humans are almost indistinguishable from other hepatitis forms except for epidemiological features and risk factors. In areas of high endemicity most patients present a short prodrome, followed by dark urine and jaundice (Aggarwal 2011). Most patients recover spontaneously, but some develop acute hepatitis and, for unknown reasons, the disease can progress into liver failure in pregnant women and immune compromised subjects. In areas with low endemicity for hepatitis E, more variable symptoms have been described, and some patients did not have typical hepatitis symptoms and/or abnormal values of liver parameters (Aggarwal and Jameel 2011). More recently, chronic infections with hepatitis E virus have been observed in patients who have received an organ transplant or immunosuppressive drugs (Kamar et al. 2011; Lhomme et al. 2012; Pas et al. 2012). Unusual symptoms including neurological sequelae have also been reported in both acute and chronic infections, sometimes representing dominant signs in the overall clinical picture (Kamar et al. 2011).

7.1 Laboratory Detection and Diagnosis in Humans

Apart from clinical symptoms, HEV infection is commonly identified in humans either indirectly by detecting a host immune response to HEV antigens, or directly by detecting the viral nucleic acid in serum samples.

Acute hepatitis E diagnosis is based on evidence of the symptoms described above, including all parameters of liver failure, together with the presence of increasing anti-HEV IgG titers and/or the presence of anti-HEV IgM and/or HEV RNA in the serum and/or stool samples (Renou et al. 2009; Pavio et al. 2010). After the incubation period, there is normally an increase of the anti-HEV IgM level for a short time, typically 2–3 weeks, followed by a more durable anti-HEV

F. M. Ruggeri et al., *Hepatitis E Virus*, SpringerBriefs in Food, Health, and Nutrition,
DOI: 10.1007/978-1-4614-7522-4_7, © Franco Maria Ruggeri, Ilaria Di Bartolo,
Marcello Trevisani, Fabio Ostanello 2013

IgG response (Huang et al. 2010). HEV specific antibodies remain detectable over a much longer period of time than does viral RNA, which enables performing a diagnosis during a longer detection window (Panda et al. 2007). Serological diagnosis is generally performed using enzyme-linked immunosorbent assay (ELISA), or rapid immune chromatographic kits (Legrand-Abravanel et al. 2009; Kamar et al. 2012) using as target antigens either recombinant HEV proteins or synthetic HEV peptides corresponding to antigen epitopes of the structural HEV proteins, such as pORF2 and pORF3 (Ma et al. 2009, 2011). Despite the existence of different HEV genotypes that can infect humans and pigs, only one serotype has been described so far (Emerson and Purcell 2003). Consequently, the same antigen of any genotype can be used to test all HEV genotypes. However, pORF2 derived recombinant antigens have proven to be more sensitive and more specific than pORF3. As suggested in several studies, antigens corresponding to capsid proteins from one particular genotype may have a better reactivity with antibodies from humans and animals infected with HEV belonging to the same genotype (Aggarwal 2012). In the absence of different serotypes, this apparent difference does not seem to have a clear justification, although it could be linked to the existence of minor diversity in some protein epitopes between different genotypes that otherwise share common neutralization epitopes (Emerson et al. 2006; Aggarwal 2012).

The antibody detection methods most frequently used are based on ELISA test platforms. Presently, several commercial ELISA kits are available for detection of both IgM and IgG antibodies, using recombinant viral antigens attached to the solid phase surface, in an indirect or double-antigen sandwich ELISA format. There are numerous examples of application of recombinant pORF2, with or without pORF3, in the diagnostic assays. Different antigens have been used, such as E. coli-expressed HEV antigens (Wang et al. 2001; Obriadina et al. 2002), and virus-like particles expressed in recombinant insect cells (Li et al. 2000; Innis et al. 2002; Jimenez de Oya et al. 2009) or in plant cells systems (Aggarwal 2012). With some modifications, these antigens have been used for detection of IgG, IgM, and IgA antibodies (Takahashi et al. 2005). IgM antibody detection, which has a relevant clinical implication, can be hampered by competition with IgG immunoglobulins, which are generally present at higher levels. To circumvent these problems and to differentiate the detection of IgM or IgG responses, a possible approach is the use of secondary antibodies directed against either the γ- or the μ-chains of immunoglobulins, respectively (Khudyakov and Kamili 2011). Several studies performed in industrialized countries have reported antibody prevalence ranging between 1 and 20 % in the normal population. Some of these values appear to be relatively high compared with the low prevalence of clinical disease in the areas considered. It still remains unclear whether the high anti-HEV levels in non-endemic areas may reflect subclinical or anicteric HEV infection, serological cross-reactivity with other agents, false positivity of serological tests, subclinical infection with possibly attenuated swine HEV or other HEV-like viruses, or a combination of more of these factors. Unfortunately, serological tests currently used differ considerably and also have varying degrees of sensitivity and specificity, thus often complicating the interpretation and comparison of results reported in some studies.

Laboratory diagnostic methods for direct diagnosis of infection include molecular techniques and, less frequently, transmission electron microscopy, that allow detection of the nucleic acid or the direct observation of virus particles, respectively. Detection of the virus has also been performed by immune electron microscopy (IEM), which actually provided the first evidence on the existence of HEV (Bradley et al. 1987), and presents the advantage of concentrating virions in the microscopy field. However, IEM remains quite insensitive compared to molecular detection methods, and requires expensive instruments and skilled personnel that make it unsuitable for clinical diagnostic routine. Direct HEV diagnosis is, therefore, preferentially performed by detection of viral nucleic acid in feces or serum samples by RT-PCR or other molecular biology based methods. The viral RNA is present in blood for about 3 weeks after the onset of symptoms, and the virus is shed with the feces for about 2 weeks. This reduces the windows of detectable viral nucleic acids to a variable short interval, requiring that the technique applied to detect nucleic acid is highly sensitive and specific (Kamar et al. 2012). The most frequently used detection methods include reverse transcription-polymerase chain reaction, using both conventional (RT-PCR) and real-time (RT-qPCR) amplification, and other approaches, used less frequently, such as transcription-loop-mediated isothermal amplification (Lan et al. 2009).

Several RT-PCR protocols have been developed that vary in the length and the position of the HEV genome segment that is amplified, as well as in their ability to detect one or more genotypes. However, most primers normally in use enable amplification of all genotype genomes, being directed at conserved sequences. Further analysis of the amplified PCR product for identification and characterization of the virus genotype is usually performed by nucleotide sequencing, by restriction with endonuclease enzymes, or other detailed investigations following molecular cloning.

Recently, great improvement of diagnostic methods has been provided by the implementation of several optimized real-time PCR protocols; this technique is not only able to quantify the virus load, but it also greatly enhances the rapidity, sensitivity, and specificity of the test.

The two real-time protocols most frequently used have been designed on the ORF2/ORF3 overlapping region, and can be suitably used for detection of HEV genotypes 1 through 4 (Jothikumar et al. 2006; Gyarmati et al. 2007).

However, a comparative study recently conducted by laboratories expert in HEV detection on behalf of the WHO showed some extent of variability in the accuracy obtained with all these methods between different laboratories, showing in particular high inter-laboratory variation in sensitivity. This study confirms the importance and need of multi-centric standardization, as well as of suitable control materials to be used as international standards, for improving diagnostic assay aimed at quantification of the HEV RNA (Kamar et al. 2012).

7.2 Laboratory Detection and Diagnosis in Swine

Except for avian HEV, in all animals investigated to date the disease is asymptomatic. In swine the persistence of maternal immunity for approximately 2 months makes it more likely that natural infection occurs at approximately 2–3 months of age (Nakai et al. 2006). Therefore, pigs of about 3–5 months of age, in which HEV infection is likely to be still active, could be preferentially subjected to direct search of the virus. Indeed, serological studies in swine have shown a higher prevalence of anti-HEV-antibodies in breeding herds (Di Bartolo et al. 2011) compared to finisher and nursery pigs (Zhang et al. 2011). Anti-HEV-IgM can be first detected in 13-week-old pigs, and also less frequently in pigs at slaughter age (Casas et al. 2011). Overall, serological testing gives a high rate of positive results especially when performed in adult animals. Diagnosis in swine is performed with the same techniques used for humans, not only when testing serum and feces, but also when analyzing tissues and organs (liver and bile) collected from slaughtered animals or from animals subjected to necropsy. Animal strains of HEV, particularly those belonging to genotype 3, are correlated genetically and antigenically to human and less to avian strains. A large part of the tests adopted for the detection of anti-HEV antibodies in animals utilize commercial human HEV-immunoassays, where the secondary antibodies are replaced with specific antisera according to the animal species concerned (Clemente-Casares et al. 2003). Alternatively, in some studies HEV-specific antibodies in animals have been detected by in-house assays developed using recombinant pORF2 or pORF3 proteins, originating from different genotypes (Mizuo et al. 2002). Abundant literature (Li et al. 2000; Peralta et al. 2009b) supports that several epitopes, especially present in pORF2, cross-react between different species and strains. A species-independent double-antigen sandwich ELISA was accordingly developed, and used to detect antibodies against HEV in swine, wild boar, and deer, resulting both sensitive and specific for the target HEV (Hu et al. 2008; Rutjes et al. 2010).

If serological testing may be still subject to some limitation related to specific antigenic sites in some HEV strains, conversely, assays established for the detection of HEV RNA have the advantage of being largely independent of genotypes as well as of the type of specimen to be tested. In fact, most protocols used for humans are also satisfactorily applied in the case of animals, spanning strains belonging to genotypes 1 through 4. Most of the primers universally applicable anneal in ORF2 (Huang et al. 2002; Mizuo et al. 2002), in the methyl-transferase (MeT) (Erker et al. 1999), or in the conserved region of the RNA-dependent RNA-polymerase (RdRp) (Zhai et al. 2006). Moreover, most conventional PCR can be combined with a nested PCR allowing one to increase the sensitivity of detection of low level viral RNA which is known to be generally low in the case of HEV, particularly in the swine. Recently, primers designed on the genotypes 1–4 were found to be suitable to detect the novel HEV strains discovered in bats and wild rats (Johne et al. 2010b; Drexler et al. 2012). Besides the overall suitability of universal primers for any type of HEV, some differences in sensitivity have, however, been shown when using two common quantitative real-time RT-PCRs on different subtypes of genotype 3 (Abravanel et al. 2012). This result remarks the necessity of preparing sets of control materials to be used as internal controls for each specific genotype. As in the case of conventional PCR tests, the real-time RT-PCR protocols used in animals and humans are the same (Jothikumar et al. 2006; Gyarmati et al. 2007), with primers and probes that can be used to detect all genotypes (1–4). Recently, a new reverse transcription loop-mediated isothermal amplification test was assessed for swine HEV detection, reporting that the sensitivity of this assay was higher than that of the conventional nested RT-PCR assay. Moreover, because of the simplicity in its performance, the assay described appeared to be suitable for use in the diagnosis of swine HEV, not only in the specialized laboratory but also under field conditions (Zhang et al. 2012).

Nucleic acid extraction can be easily and efficiently performed, using samples of feces, serum, or a wide range of tissues. Virus detection for risk assessment in food is generally performed in tissues such as liver, because the liver is normally used for producing sausage in many countries, but also in the intestine and lymph nodes (de Deus et al. 2007). Several studies have proven that the highest sensitivity in HEV detection is achieved using swine samples, such as liver, mesenteric lymph nodes, and bile (de Deus et al. 2007) that cannot be obtained from live animals, followed by feces that, together with sera, represents the easiest to collect and most commonly used type of sample in field studies aimed at viral RNA detection. Detection of HEV genome was also reported in the bladder, tonsils, and urine samples of infected animals (Bouwknegt et al. 2009; Leblanc et al. 2010). Viral loads of 10^3–10^7 genome copies/g were estimated in positive liver and bile samples (Leblanc et al. 2010). In addition to RT-PCR, in situ hybridization and immunohistochemistry can be used in testing of organs. These techniques enable a precise localization of the virus in the tissues and infected cells, and offer a reliable tool for investigating the possible correlations between the virus replication and local lesions and to identify virus replication sites (Lee et al. 2009b). Detecting the virus in animal organs, particularly liver, is not only important to assess the possible zoonotic risks associated with the consumption of a specific infected food, but has also opened new possibilities in the understanding of the pathogenesis of HEV infection.

7.3 Virus Isolation in Cell Cultures

Since the discovery of hepatitis E virus, several attempts have been made to grow the virus in vitro, most of which were fully unsuccessful or at most have permitted limited virus replication and low titers of progeny viruses. In general, growing the virus in cultured cells is extremely troublesome. Two main factors can hamper HEV growth in vitro: (i) the normally low viral load in test samples: a higher titer warrants a higher probability to get replication in a short time; (ii) the virus integrity in the samples used for the inoculums: specimens should be put in cell cultureno later

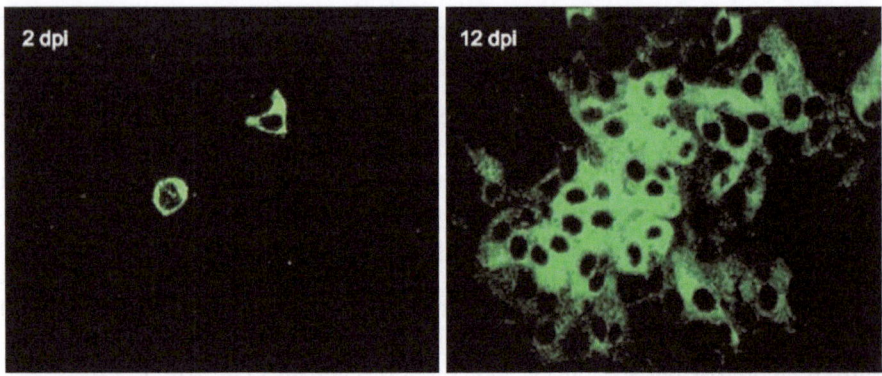

Fig. 7.1 Immunofluorescent staining of HEV inoculated PCL/PRF/5 cells by an anti-ORF2 MAb. Reproduced from Okamoto et al. (2011), with permission of Elsevier

than 6 months after collection, otherwise isolation is less likely to succeed despite sample storage at very low temperature (Huang et al. 1999). Human lung cancer cells A549 were first used for successful isolation of a human HEV strain (Huang et al. 1999). In the last few years, several papers reported attempts to cultivate different viral strains in vitro using cell lines of either human or animal origin (Okamoto 2013). Only more recently the first efficient in vitro cultivation of a genotype 3 strain in both a hepatocarcinoma (PLC/PRF/5) and a human lung cancer (A549) cell line has been reported. The experiments performed clearly demonstrated that both the initial day of HEV appearance in the infected cell culture supernatant and its viral load (determined by qPCR, and expressed in copy numbers) were largely dependent on the original titer of the virus used for inoculums (Okamoto 2007). HEV could be first detected in the culture supernatant 8 days post-inoculation; however, it reached a concentration of 10^8 genome copies/ml only after several weeks (40–50 days) (Okamoto 2011). The A549 and PLC/PRF/5 cell lines have been used in several studies, demonstrating efficient replication of both HEV genotypes 3 and 4 of both human and animal origin, including strains from swine, wild boar and deer. Moreover, the virus progeny was infectious, as established by repeated passages in the same or other cell lines (Okamoto 2011; Takahashi et al. 2012). The PLC/PRF/5 cells used for the monolayer cultures were also successfully used in a 3D system (Berto et al. 2013b), using special cell incubation conditions supported on microspheres. In addition, a recent study has further increased the number of cell lines permissive to HEV growth, describing successful viral replication in the human hepatoma-derived cell line HepaRG, and the porcine embryonic stem cell-derived cell line PICM-19 (Rogee et al. 2012).

Development of cell cultures is expected to give an important contribution to the diagnosis of HEV infections, and in the future to the evaluation of antiviral treatments. Various HEV strains of genotypes 1, 3, and 4, obtained from serum samples of patients with hepatitis E, have been successfully replicated in vitro in both PCL/PRF/5 and A549 cell lines. Since HEV represents an emerging topic in donors' blood safety, because of potential transmissibility of the virus through transfusions (Okamoto 2013), the use of cell cultures may be critical to assess the actual risks associated with blood donations, adding direct proof of infectivity to the possible finding of viral genomic RNA that is the only possible parameter obtained at present.

Some studies conducted using cells culture have demonstrated that pork liver sausage and swine liver sold as food can contain infectious HEV (Takahashi et al. 2012; Berto et al. 2013a) identifying the risk for foodborne HEV transmission. However, the use of HEV cultivation in vitro to assess the risk in food items is still too cumbersome for being presently considered suitable for application in the practice, calling for further studies in order to develop reliable and handy protocols (Fig. 7.1).

Chapter 8
Hepatitis E Virus: Food Related Issues

In assessing the risk related to the presence of HEV in food some facts need to be evaluated.

Hepatitis E occurs in both epidemic and sporadic–endemic forms. Insufficient drinking water treatment and low standards of sanitation have been implicated in major outbreaks in developing countries. Major waterborne epidemics have occurred in Asia and North and East Africa, but the infective dose is not yet known. Experimental data from non-human primates indicate that the severity of infection with HEV is proportional to the infectious dose (Tsarev et al. 1994). The estimated orally ingested HEV dose at which the probability of infection equals 50 % in pigs is 1.4×10^6 HEV genome copies (Bouwknegt et al. 2011). However, the rate of clinical disease following infection depends on susceptibility and in humans, previous exposure (immunity), pregnancy and the presence of other diseases play a significant role. Person-to-person spread has been documented and the fecal–oral route denotes problems in hygiene management in some cases (Said et al. 2009), and anecdotal cases of transmission by solid-organ transplant and blood transfusion were reported (Matsubayashi et al. 2008; Legrand-Abravanel et al. 2010). The disease has an incubation period of 15–60 days, usually is mild and resolves in 2 weeks, leaving no sequelae. The disease is most often seen in young to middle aged adults (15–40 years old). The fatality rate is 0.1–1 % except in pregnant women. This group is reported to have a fatality rate approaching 20 %. Association of clinical disease and alcohol consumption was observed in groups of individuals exposed during cruise ship. Clinical cases of hepatitis E have been reported sporadically from industrialized countries like the USA, Japan, Taiwan, France (Hsieh et al. 1999; Clemente-Casares et al. 2003; Mizuo et al. 2005) despite the relatively high level of HEV antibody prevalence in healthy individuals (Meng et al. 2002; Mansuy et al. 2011). In individuals exposed to recent HEV infection, as shown by the presence of a seroconversion peak, who did not have symptomatic hepatitis, also did not have abnormal results of liver function tests (Said et al. 2009). Subclinical infection with HEV genotypes 3 and 4 may be responsible for the high prevalence of anti-HEV in industrialized countries (Purcell and Emerson 2010).

F. M. Ruggeri et al., *Hepatitis E Virus*, SpringerBriefs in Food, Health, and Nutrition, DOI: 10.1007/978-1-4614-7522-4_8, © Franco Maria Ruggeri, Ilaria Di Bartolo, Marcello Trevisani, Fabio Ostanello 2013

Evidence of the presence of HEV in pigs, wild boar, and water is well documented in many countries and the high prevalence in farmed animals and the environment is also associated with a relevant (even high) prevalence of infection in humans (Banks et al. 2004b; Zheng et al. 2006; Caprioli et al. 2007a; Feagins et al. 2007b; Schielke et al. 2009). Epidemiological studies have shown that human and swine HEV isolates are genetically closely related (Lu et al. 2006; Zheng et al. 2006). Rodents have also been suggested to be a reservoir for HEV (Favorov et al. 2000; Easterbrook et al. 2007; Vasickova et al. 2007). The presence of HEV RNA has been shown in 14 % of the pig fecal samples and 6 % of commercial pig livers in a study from the Netherlands (Rutjes et al. 2009).

A recent study carried out by Kanai and colleagues (Kanai et al. 2012) showed that 28.6 % of rats living close to a farm housing infected pigs were seropositive and detection of HEV RNA was possible in 17.8 % of them suggesting a possible role of rats in spreading of the infections. However, the role of rats in human infection remains controversial. Hepatitis E virus genotype 3 was detected in wild rats living far from rural or urban areas in the USA (Lack et al. 2012), whereas other HEV strains detected in rats seem to represent a novel genotype (Johne et al. 2012).

Hepatitis E has been detected in wastewater resulting from gut processing at slaughterhouses (Rutjes et al. 2009) and in pig slurry stores (McCreary et al. 2008). HEV was found in 4.76 % of lettuce sampled at primary production level and 3.2 % samples at the point of sale (6–23 mean PCR detectable units) and 5 % irrigation water in an EU survey (Kokkinos et al. 2012).

The survival of HEV genotype 1 in soil was analyzed (Parashar et al. 2011). The authors assessed that 4.9 % of the 403 samples collected from four locations along the Mutha river (India) were positive for HEV RNA and they observed that sewage discharges play a significant role in maintaining this contamination. They also artificially spiked soil samples (0.5 g) collected from various locations with 6.47×10^7 RNA copies of the virus in order to test the stability of the genome and observed a sudden decrease after 3 weeks, reaching a plateau during weeks 3–8.

Hepatitis E virus was also detected in shellfish, such as mussels and oysters farmed in Scotland (Crossan et al. 2012). Overall, 92 % of the productive sites located on the west coast and 55 % of those on the east coast were positive. The prevalence of contamination, which was shown by HEV RNA detection, was very high (range 3.73–5.2 $\log_{10}$ IU/ml) in a site that was near a purification plant that processes water released by a local swine slaughterhouse and meat processing plant. HEV was also detected in bivalve shellfish (*Corbicula japonica*) purchased from a fish market in Japan (Li et al. 2007).

Very often the serological tests show the presence of infections not associated with disease in both animals and humans. The higher prevalence is observed in farms housing pigs 2–4 months of age (Meng et al. 1997; Takahashi et al. 2003; Rutjes et al. 2009) whereas it is lower in pigs at ages of 6–8 months, which is the age at slaughter for most countries (e.g. EU, USA, AUS). Recent studies have shown, however, that infection can occur at all ages, from weaners to fatteners and that pigs close to the slaughter age can still be HEV infected (Berto et al. 2012a).

Pig meat may, therefore, be a vehicle of infection for the consumers and infection can be also transmitted more frequently to the persons who more commonly are in contact with the infected animals, like butchers, pig farmers, and workers at the slaughterhouses. The presence of HEV in muscle samples of swine was observed up to 4 weeks after the onset of fecal shedding and the time of infection has been considered important for the presence of HEV in pork meat with late infection during the fattening period having a major impact on the presence of HEV-positive muscle in retail stores (Bouwknegt et al. 2009). Consumption of raw pig meat was associated with cases of hepatitis E in the Netherlands (Melenhorst et al. 2007). At pig slaughterhouses, tools (knives) and surfaces (belt and floor) were found to be positive for HEV RNA (Di Bartolo et al. 2012). The contamination of tools and environmental samples (knife and slicer) was also proven at retail shops (Berto et al. 2012b). Hygienic practices may play an important role in determining the higher level of contamination and, therefore, the probability of infection. Consumption of pork sausages prepared from house-slaughtered meat was the suspected cause of a case observed in Hungary in a 60-year-old man (Reuter et al. 2006).

The risk related to consumption of raw pig organ meat is well documented. Contamination was detected in 9 out of the 43 (20.9 %) liver samples taken from adult swine belonging to an infected herd showing high titers up to $1.0–9.9 \times 10^7$ genome copies/g (Leblanc et al. 2010). The presence of HEV was detected in a relevant percentage of samples of hepatic lymph nodes, bile, bladders, and feces, showing that cross-contamination of meat is an issue that must be considered. The presence of HEV was also observed in livers from wild boar with an average detection rate of 14.9 % in Germany (Schielke et al. 2009).

Sequences of HEV that are strictly related genetically with those recovered from human cases were detected in samples of raw smoked liver sausages (i.e. *figatellu*) during a case–control study that was carried out in Corsica (France) in 2010 (Colson et al. 2010). The titer of HEV detected in the *figatellu* purchased from a local market was $10^3–10^6$ genome copies per slice, although the viability/ infectivity of the virus was not tested.

Thermal stability of HEV was analyzed in monkeys (Emerson et al. 2005) by using fecal suspensions containing a 50 % monkey infectious dose (MID50) of $10^{6.5}/0.5$ ml of 10 % fecal suspension. Differences were observed between the two strains used (genotype 1 and 2). The most resistant strain showed intact infectivity in cultured cells (HepG2/C3A) after treatment at 56 °C and reduction of 80 % of its infectivity after a treatment at 60 °C for 30 min. Feagins and colleagues (Feagins et al. 2008) have assessed instead the thermal stability of HEV by cooking liver blocks (size 0.5–1 cm^2) in boiling water for 5 min or by stir-frying until the internal temperature reached at least 71 °C. The infectivity of cooked liver was assessed by inoculating the treated sample homogenates to groups of healthy pigs (that were negative at HEV antibodies testing) intravenously. Both treatments showed a complete inactivation of HEV. No seroconversion, viremia, or fecal virus shedding was observed in any of the inoculated pigs. In order to better understand the possible inactivation rate in the conditions that are used during industrial processing,

(Barnaud et al. 2012) pâté-like preparations were produced (30 % infected liver; 48 % fat and brine) and different time/temperature combinations were used, ranging between 62 and 71 °C and 5–20 min. The samples used for the thermal treatments had a weight of 25 g and thickness of 2 mm. The residual infectivity was tested by experimentally inoculate homogenates intravenously to group of pigs and testing the seroconversion and viral extraction in feces. Pâté cooked at 62 °C for 120 min still allow the survival of infective virus, whereas the rate of infection was reduced in the group that received pâté treated at 68 °C for up to 20 min and 71 °C up to 10 min. In pâté-like preparations, therefore, only treatments at 71 °C for 20 min allowed a complete loss of residual infectivity. Guidelines for production of liver sausages or pâté indicate that a core temperature to be reached is at least 74 °C (Heinz and Hautzinger 2007) but many receipts advice 78 °C.

Association between shellfish consumption and HEV infection was suspected in outbreaks (Said et al. 2009). Consumption habits play a fundamental role for the risk of HEV transmission. For example, shellfish can be consumed after appropriate cooking, such as for Yamato-Shijimi, which is generally eaten as an ingredient in hot miso soup in Japan, where the heat usually reaches 100 °C for nearly 10 min or not, as is the case of oysters that are consumed raw or mussels that are generally rapidly steam cooked until their shells open, reaching an internal temperature of approximately 60 °C for few a minutes.

No risks are associated with shellfish or crustaceans after boiling or steaming under pressure to achieve a minimum temperature in the product of 90 °C for not less than 90 s or its equivalence. Improper handling can, however, result in cross-contamination of food after a lethal thermal treatment.

The following considerations can be done on the basis of the facts described above. Pig and wild boar are carriers of HEV and very high titers of virus can be found in effluents from pig farms and slaughterhouses.

The temperature of scalding water and the time used for allowing hair removal (normally 58 °C for 6 min or 63 °C for 2 min) are not effective to inactivate HEV, but the importance of cleanliness and pre-scald treatment (e.g. hot water washing or passage in separated vats) should not be underestimated.

HEV can be spread during the cutting of liver and meat on industrial premises, requiring cross-contamination control measures. No data are available concerning the effectiveness of the biocides used for sanitation in the meat industry. Hepatitis E virus does not have a lipid envelope and control of surface contamination with quaternary ammonium compounds or other tensioactive compounds is not a good choice.

Contaminated water may carry infective particles to vegetables and shellfish farms. Outbreaks related to water contamination have shown that HEV does not respond to chlorination and dissolved, colloidal, and solid organic matter protect the virus from inactivation. Therefore, treatment of slaughterhouses and pig farms wastewater should be improved to prevent environmental safety issues.

Rats may play a critical role in HEV outbreaks, considering their negative effects on environmental hygiene and food sanitation. Water and rodents may also be a link between rural environment and wildlife (e.g. wild boar hunted close to urban areas). The widespread presence of HEV in rats determines concerns for food safety.

Slaughterhouse workers, pig farmers, and butchers are frequently exposed to HEV. Human infection seems to be related, however, to the exposure to high titers of infective HEV (i.e. not inactivated by thermal treatments). Mechanisms of inactivation of HEV by UV radiation and ozone (i.e. the treatment most commonly used for the treatment of recycled water in shellfish depuration) need to be investigated.

The frequent exposure in humans only seldom results in symptomatic infection. Most commonly they show seroconversion without signs of hepatitis, nor are they detected by liver function tests (Said et al. 2009). Some evidence indicated association with alcohol consumption and symptomatic cases, but also higher disease fatality rate in pregnant women. It is worth investigating the human factors that may affect the outcomes, such as immunization by previous HEV infection or other hepatic diseases.

Chapter 9
Conclusions

Epidemiological and virological studies conducted in the last few years have clearly demonstrated that hepatitis E should be considered an emerging zoonosis. Swine appears to represent the major animal reservoir for the virus, and the large global relevance of the pork food chain creates public health concerns for the future. HEV infection can be transmitted through food by the ingestion of infected meat products (Matsuda et al. 2003; Tei et al. 2003; Yazaki et al. 2003; Banks et al. 2004c; Tamada et al. 2004). The risk associated with this mode of transmission is, however, reduced in developed countries, because pork is mostly consumed cooked, and viruses are inactivated during the process of cooking. In addition, HEV is prevalent particularly in swine below the age of 5 months (i.e. below the slaughter age, in most countries). However, the possibility of cross-contamination between raw meat products and the risk of virus spread in the environment through manure from pig farms, with the consequent possible contamination of vegetables and drinking or bathing water, should also be taken into consideration. Contamination of water, as is the case with hepatitis A virus, could also lead to the contamination of filtering shellfish, thereby further compounding public health risks.

Another possible route of transmission of hepatitis E virus to humans is by direct contact with infected animals. In this case, people such as farmers, workers attending the animals, and veterinarians who work in contact with pigs during the viremic period or when virus is excreted in the feces may be at greater risk of infection (Meng et al. 1997; Drobeniuc et al. 2001; Withers et al. 2002; Yazaki et al. 2003). Furthermore, for these population categories, the possibility of infection by indirect contact with instruments and tools contaminated with infected feces cannot be ruled out. The knowledge of these risks should thus encourage recourse to hygiene and biosecurity procedures that may help avoid or minimize the chances of infection. The discovery of HEV may also pose risks during the practice of xenotransplantation, which has become a possible solution to the problem of a shortage of organ donors for transplantations. This risk is not limited to HEV. In fact, otherwise non-pathogenic viruses of pigs belonging to different families might become pathogenic for humans after xenotransplantation as a result of

F. M. Ruggeri et al., *Hepatitis E Virus*, SpringerBriefs in Food, Health, and Nutrition, DOI: 10.1007/978-1-4614-7522-4_9, © Franco Maria Ruggeri, Ilaria Di Bartolo, Marcello Trevisani, Fabio Ostanello 2013

crossing the species barrier, recombination or adaptation in immune-compromised xenotransplantation in addition to normal transplantation recipients (Meng et al. 2002; Pei and Yoo 2002; Razonable 2011; Pas et al. 2012).

The enzootic nature of swine HEV infection in pigs in many countries, together with its ability to cross the species barrier, raise concerns with regards to the possibility of zoonotic transmission, and food and environmental safety. Nevertheless, many veterinary aspects of infection are not yet known. Knowledge is still limited with respect to the genetic correlation between different animal and human strains of HEV. The natural history of infection in pigs also requires further study, as does the economic impact of the disease on pig production. The host range of the infection is not fully known, and the cases associated with the ingestion of uncooked meat from wild boar and deer in Japan indicate that the role of wild animals, as well as swine and ruminants, should be further considered in the epidemiology of the disease.

Despite the obvious health implications of this emerging zoonosis, information on the significance of HEV circulation in swine herds, and in other animals in developed countries including Europe is still insufficient. In countries like Spain, the Netherlands, and the UK, where larger epidemiological studies have been conducted, it has been demonstrated that the virus is circulating actively in the pig herds. More recently, similar results have been achieved in Italy, and other European countries, highlighting the wider scale of the phenomenon. If the possibility of zoonotic transmission of the infection is considered, it is clear that the higher the prevalence in animals, the greater the risk of transmission will be to humans.

References

Abravanel F, Sandres-Saune K, Lhomme S, Dubois M, Mansuy JM, Izopet J (2012) Genotype 3 diversity and quantification of hepatitis E virus RNA. J Clin Microbiol 50:897–902

Adlhoch C, Wolf A, Meisel H, Kaiser M, Ellerbrok H, Pauli G (2009) High HEV presence in four different wild boar populations in East and West Germany. Vet Microbiol 139:270–278

Aggarwal R (2011) Clinical presentation of hepatitis E. Virus Res 161:15–22

Aggarwal R (2012) Diagnosis of hepatitis E. Nat Rev Gastroenterol Hepatol 10:24–33

Aggarwal R, Jameel S (2011) Hepatitis E. Hepatology 54:2218–2226

Aggarwal R, Krawczynski K (2000) Hepatitis E: an overview and recent advances in clinical and laboratory research. J Gastroenterol Hepatol 15:9–20

Aggarwal R, Naik S (2009) Epidemiology of hepatitis E: current status. J Gastroenterol Hepatol 24:1484–1493

Aggarwal R, Shukla R, Jameel S, Agrawal S, Puri P, Gupta VK et al (2007) T-cell epitope mapping of ORF2 and ORF3 proteins of human hepatitis E virus. J Viral Hepat 14:283–292

Agrawal S, Gupta D, Panda SK (2001) The 3' end of hepatitis E virus (HEV) genome binds specifically to the viral RNA-dependent RNA polymerase (RdRp). Virology 282:87–101

Agrawal V, Goel A, Rawat A, Naik S, Aggarwal R (2012) Histological and immunohistochemical features in fatal acute fulminant hepatitis E. Indian J Pathol Microbiol 55:22–27

Agunos AC, Yoo D, Youssef SA, Ran D, Binnington B, Hunter DB (2006) Avian hepatitis E virus in an outbreak of hepatitis–splenomegaly syndrome and fatty liver haemorrhage syndrome in two flaxseed-fed layer flocks in Ontario. Avian Pathol 35:404–412

Ahmad I, Holla RP, Jameel S (2011) Molecular virology of hepatitis E virus. Virus Res 161:47–58

Arankalle VA, Chobe LP, Chadha MS (2006) Type-IV Indian swine HEV infects rhesus monkeys. J Viral Hepat 13:742–745

Arankalle VA, Joshi MV, Kulkarni AM, Gandhe SS, Chobe LP, Rautmare SS et al (2001) Prevalence of anti-hepatitis E virus antibodies in different Indian animal species. J Viral Hepat 8:223–227

Backer JA, Berto A, McCreary C, Martelli F, van der Poel WH (2012) Transmission dynamics of hepatitis E virus in pigs: estimation from field data and effect of vaccination. Epidemics 4:86–92

Balayan MS, Andjaparidze AG, Savinskaya SS, Ketiladze ES, Braginsky DM, Savinov AP et al (1983) Evidence for a virus in non-A, non-B hepatitis transmitted via the fecal-oral route. Intervirology 20:23–31

Banks M, Bendall R, Grierson S, Heath G, Mitchell J, Dalton H (2004) Human and porcine hepatitis E virus strains, United Kingdom. Emerg Infect Dis 10:953–955

Banks M, Heath GS, Grierson SS, King DP, Gresham A, Girones R et al (2004) Evidence for the presence of hepatitis E virus in pigs in the United Kingdom. Vet Rec 154:223–227

Banks M, Heath GS, Grierson SS, King DP, Gresham A, Girones R et al (2004) Evidence for the presence of hepatitis E virus in pigs in the United Kingdom. Vet Rec 154:223–227

Banyai K, Toth AG, Ivanics E, Glavits R, Szentpali-Gavaller K, Dan A (2012) Putative novel genotype of avian hepatitis E virus, Hungary, 2010. Emerg Infect Dis 18:1365–1368

Barnaud E, Rogee S, Garry P, Rose N, Pavio N (2012) Thermal inactivation of infectious hepatitis E virus in experimentally contaminated food. Appl Environ Microbiol 78:5153–5159

Batts W, Yun S, Hedrick R, Winton J (2011) A novel member of the family Hepeviridae from cutthroat trout (Oncorhynchus clarkii). Virus Res 158:116–123

Berto A, Backer JA, Mesquita JR, Nascimento MS, Banks M, Martelli F et al (2012) Prevalence and transmission of hepatitis E virus in domestic swine populations in different European countries. BMC Res Notes 5:190

Berto A, Grierson S, van der Hakze Honing R, Martelli F, Johne R, Reetz J et al (2013) Hepatitis e virus in pork liver sausage, france. Emerg Infect Dis 19:264–266

Berto A, Martelli F, Grierson S, Banks M (2012) Hepatitis E virus in pork food chain, United Kingdom, 2009–2010. Emerg Infect Dis 18:1358–1360

Berto A, Mesquita JR, van der Hakze Honing R, Nascimento MS, van der Poel WH (2012) Detection and characterization of hepatitis e virus in domestic pigs of different ages in Portugal. Zoonoses Public Health 59:477–481

Berto A, Van der Poel WH, Van der Hakze Honing R, Martelli F, La Ragione RM, Inglese N et al (2013) Replication of hepatitis E virus in three-dimensional cell culture. J Virol Methods 187:327–332

Bile K, Isse A, Mohamud O, Allebeck P, Nilsson L, Norder H et al (1994) Contrasting roles of rivers and wells as sources of drinking water on attack and fatality rates in a hepatitis E epidemic in Somalia. Am J Trop Med Hyg 51:466–474

Bilic I, Jaskulska B, Basic A, Morrow CJ, Hess M (2009) Sequence analysis and comparison of avian hepatitis E viruses from Australia and Europe indicate the existence of different genotypes. J Gen Virol 90:863–873

Billam P, Huang FF, Sun ZF, Pierson FW, Duncan RB, Elvinger F et al (2005) Systematic pathogenesis and replication of avian hepatitis E virus in specific-pathogen-free adult chickens. J Virol 79:3429–3437

Boadella M, Ruiz-Fons JF, Vicente J, Martin M, Segales J, Gortazar C (2012) Seroprevalence evolution of selected pathogens in iberian wild boar. Transbound Emerg Dis 59:395–404

Boccia D, Guthmann JP, Klovstad H, Hamid N, Tatay M, Ciglenecki I et al (2006) High mortality associated with an outbreak of hepatitis E among displaced persons in Darfur, Sudan. Clin Infect Dis 42:1679–1684

Bouquet J, Cherel P, Pavio N (2012) Genetic characterization and codon usage bias of full-length Hepatitis E virus sequences shed new lights on genotypic distribution, host restriction and genome evolution. Infect Genet Evol 12:1842–1853

Bouquet J, Cheval J, Rogee S, Pavio N, Eloit M (2012) Identical consensus sequence and conserved genomic polymorphism of hepatitis E virus during controlled interspecies transmission. J Virol 86:6238–6245

Bouwknegt M, Frankena K, Rutjes SA, Wellenberg GJ, Husman AMDR, van der Poel WHM et al (2008) Estimation of hepatitis E virus transmission among pigs due to contact-exposure. Vet Res 39

Bouwknegt M, Lodder-Verschoor F, van der Poel WH, Rutjes SA, de Roda Husman AM (2007) Hepatitis E virus RNA in commercial porcine livers in The Netherlands. J Food Prot 70:2889–2895

Bouwknegt M, Rutjes SA, Reusken CB, Stockhofe-Zurwieden N, Frankena K, de Jong MC et al (2009) The course of hepatitis E virus infection in pigs after contact-infection and intravenous inoculation. BMC Vet Res 5:7

Bouwknegt M, Teunis PF, Frankena K, de Jong MC, de Roda Husman AM (2011) Estimation of the likelihood of fecal-oral HEV transmission among pigs. Risk Anal 31:940–950

Boxall E, Herborn A, Kochethu G, Pratt G, Adams D, Ijaz S et al (2006) Transfusion-transmitted hepatitis E in a 'nonhyperendemic' country. Transfus Med 16:79–83

Bradley DW, Krawczynski K, Cook EH Jr, McCaustland KA, Humphrey CD, Spelbring JE et al (1987) Enterically transmitted non-A, non-B hepatitis: serial passage of disease in cynomolgus macaques and tamarins and recovery of disease-associated 27- to 34-nm viruslike particles. Proc Natl Acad Sci USA 84:6277–6281

Breum SO, Hjulsager CK, de Deus N, Segales J, Larsen LE (2010) Hepatitis E virus is highly prevalent in the Danish pig population. Vet Microbiol 146:144–149

Cacopardo B, Russo R, Preiser W, Benanti F, Brancati G, Nunnari A (1997) Acute hepatitis E in Catania (eastern Sicily) 1980–1994. The role of hepatitis E virus. Infection 25:313–316

Caprioli A, Martelli F, Ostanello F, Di Bartolo I, Ruggeri FM, Del Chiaro L et al (2007) Detection of hepatitis E virus in Italian pig herds. Vet Rec 161:422–423

Caprioli A, Martelli F, Ostanello F, Di Bartolo I, Ruggeri FM, Del Chiaro L et al (2007) Detection of hepatitis E virus in Italian pig herds. Vet Rec 161:422–423

Caprioli A, Ostanello F, Martelli F (2005) Hepatitis E virus: an emerging zoonotic agent. Vet Ital 41:113–127

Caron M, Enouf V, Than SC, Dellamonica L, Buisson Y, Nicand E (2006) Identification of genotype 1 hepatitis E virus in samples from swine in Cambodia. J Clin Microbiol 44:3440–3442

Carpentier A, Chaussade H, Rigaud E, Rodriguez J, Berthault C, Boue F et al (2012) High hepatitis E virus seroprevalence in forestry workers and in wild boars in France. J Clin Microbiol 50:2888–2893

Casas M, Cortes R, Pina S, Peralta B, Allepuz A, Cortey M et al (2011) Longitudinal study of hepatitis E virus infection in Spanish farrow-to-finish swine herds. Vet Microbiol 148:27–34

Chandler JD, Riddell MA, Li F, Love RJ, Anderson DA (1999) Serological evidence for swine hepatitis E virus infection in Australian pig herds. Vet Microbiol 68:95–105

Chang Y, Wang L, Geng J, Zhu Y, Fu H, Ren F et al (2009) Zoonotic risk of hepatitis E virus (HEV): A study of HEV infection in animals and humans in suburbs of Beijing. Hepatol Res 39:1153–1158

Cho JG, Dee SA (2006) Porcine reproductive and respiratory syndrome virus. Theriogenology 66:655–662

Choi C, Chae C (2003) Localization of swine hepatitis E virus in liver and extrahepatic tissues from naturally infected pigs by in situ hybridization. J Hepatol 38:827–832

Choi IS, Kwon HJ, Shin NR, Yoo HS (2003) Identification of swine hepatitis E virus (HEV) and prevalence of anti-HEV antibodies in swine and human populations in Korea. J Clin Microbiol 41:3602–3608

Cintron-Arias A, Castillo-Chavez C, Bettencourt LM, Lloyd AL, Banks HT (2009) The estimation of the effective reproductive number from disease outbreak data. Math Biosci Eng 6:261–282

Clayson ET, Innis BL, Myint KS, Narupiti S, Vaughn DW, Giri S et al (1995) Detection of hepatitis E virus infections among domestic swine in the Kathmandu Valley of Nepal. Am J Trop Med Hyg 53:228–232

Clemente-Casares P, Pina S, Buti M, Jardi R, MartIn M, Bofill-Mas S et al (2003) Hepatitis E virus epidemiology in industrialized countries. Emerg Infect Dis 9:448–454

Colson P, Borentain P, Queyriaux B, Kaba M, Moal V, Gallian P et al (2010) Pig liver sausage as a source of hepatitis E virus transmission to humans. J Infect Dis 202:825–834

Colson P, Coze C, Gallian P, Henry M, De Micco P, Tamalet C (2007) Transfusion-associated hepatitis E, France. Emerg Infect Dis 13:648–649

Colson P, Swiader L, Motte A, Ferretti A, Borentain P, Gerolami R (2012) Circulation of almost genetically identical hepatitis E virus of genotype 4 in France. J Clin Virol 55:181–183

Conlan JV, Jarman RG, Vongxay K, Chinnawirotpisan P, Melendrez MC, Fenwick S et al (2011) Hepatitis E virus is prevalent in the pig population of Lao People's Democratic Republic and evidence exists for homogeneity with Chinese Genotype 4 human isolates. Infect Genet Evol 11:1306–1311

Cooper K, Huang FF, Batista L, Rayo CD, Bezanilla JC, Toth TE et al (2005) Identification of genotype 3 hepatitis E virus (HEV) in serum and fecal samples from pigs in Thailand and Mexico, where genotype 1 and 2 HEV strains are prevalent in the respective human populations. J Clin Microbiol 43:1684–1688

Cossaboom CM, Cordoba L, Sanford BJ, Pineyro P, Kenney SP, Dryman BA et al (2012) Cross-species infection of pigs with a novel rabbit, but not rat, strain of hepatitis E virus isolated in the United States. J Gen Virol 93:1687–1695

Crossan C, Baker PJ, Craft J, Takeuchi Y, Dalton HR, Scobie L (2012) Hepatitis E virus genotype 3 in shellfish, United Kingdom. Emerg Infect Dis 18:2085–2087

Dalton HR, Fellows HJ, Gane EJ, Wong P, Gerred S, Schroeder B et al (2007) Hepatitis E in new zealand. J Gastroenterol Hepatol 22:1236–1240

Dalton HR, Thurairajah PH, Fellows HJ, Hussaini HS, Mitchell J, Bendall R et al (2007) Autochthonous hepatitis E in southwest England. J Viral Hepat 14:304–9

Dawson GJ, Mushahwar IK, Chau KH, Gitnick GL (1992) Detection of long-lasting antibody to hepatitis E virus in a US traveller to Pakistan. Lancet 340:426–427

de Deus N, Casas M, Peralta B, Nofrarias M, Pina S, Martin M et al (2008) Hepatitis E virus infection dynamics and organic distribution in naturally infected pigs in a farrow-to-finish farm. Vet Microbiol 132:19–28

de Deus N, Peralta B, Pina S, Allepuz A, Mateu E, Vidal D et al (2008) Epidemiological study of hepatitis E virus infection in European wild boars (Sus scrofa) in Spain. Vet Microbiol 129:163–170

de Deus N, Seminati C, Pina S, Mateu E, Martin M, Segales J (2007) Detection of hepatitis E virus in liver, mesenteric lymph node, serum, bile and faeces of naturally infected pigs affected by different pathological conditions. Vet Microbiol 119:105–114

de la Caridad Montalvo Villalba M, Owot JC, Corrreia B, Corredor MB, Flaquet PP, Frometa SS et al (2013) Hepatitis E virus genotype 3 in humans and swine. Cuba. Infect Genet Evol 14C:335–339

Dell'Amico MC, Cavallo A, Gonzales JL, Bonelli SI, Valda Y, Pieri A et al (2011) Hepatitis E virus genotype 3 in humans and Swine, Bolivia. Emerg Infect Dis 17:1488–1490

Di Bartolo I, Diez-Valcarce M, Vasickova P, Kralik P, Hernandez M, Angeloni G et al (2012) Hepatitis E virus in pork production chain in Czech Republic, Italy, and Spain, 2010. Emerg Infect Dis 18:1282–1289

Di Bartolo I, Martelli F, Inglese N, Pourshaban M, Caprioli A, Ostanello F et al (2008) Widespread diffusion of genotype 3 hepatitis E virus among farming swine in Northern Italy. Vet Microbiol 132:47–55

Di Bartolo I, Ponterio E, Castellini L, Ostanello F, Ruggeri FM (2011) Viral and antibody HEV prevalence in swine at slaughterhouse in Italy. Vet Microbiol 149:330–338

Dong C, Meng J, Dai X, Liang JH, Feagins AR, Meng XJ et al (2011) Restricted enzooticity of hepatitis E virus genotypes 1 to 4 in the United States. J Clin Microbiol 49:4164–4172

dos Santos DR, de Paula VS, de Oliveira JM, Marchevsky RS, Pinto MA (2011) Hepatitis E virus in swine and effluent samples from slaughterhouses in Brazil. Vet Microbiol 149:236–241

dos Santos DRL, Vitral CL, de Paula VS, Marchevsky RS, Lopes JF, Gaspar AMC et al (2009) Serological and molecular evidence of hepatitis E virus in swine in Brazil. Vet J 182:474–480

Dremsek P, Wenzel JJ, Johne R, Ziller M, Hofmann J, Groschup MH et al (2012) Seroprevalence study in forestry workers from eastern Germany using novel genotype 3- and rat hepatitis E virus-specific immunoglobulin G ELISAs. Med Microbiol Immunol 201:189–200

Drexler JF, Seelen A, Corman VM, Fumie Tateno A, Cottontail V, Melim Zerbinati R et al (2012) Bats worldwide carry hepatitis E virus-related viruses that form a putative novel genus within the family Hepeviridae. J Virol 86:9134–9147

Drobeniuc J, Favorov MO, Shapiro CN, Bell BP, Mast EE, Dadu A et al (2001) Hepatitis E virus antibody prevalence among persons who work with swine. J Infect Dis 184:1594–1597

Easterbrook JD, Kaplan JB, Vanasco NB, Reeves WK, Purcell RH, Kosoy MY et al (2007) A survey of zoonotic pathogens carried by Norway rats in Baltimore, Maryland, USA. Epidemiol Infect 135:1192–1199

Ellis J, Clark E, Haines D, West K, Krakowka S, Kennedy S et al (2004) Porcine circovirus-2 and concurrent infections in the field. Vet Microbiol 98:159–163

Emerson SU, Arankalle VA, Purcell RH (2005) Thermal stability of hepatitis E virus. J Infect Dis 192:930–933

Emerson SU, Clemente-Casares P, Moiduddin N, Arankalle VA, Torian U, Purcell RH (2006) Putative neutralization epitopes and broad cross-genotype neutralization of Hepatitis E virus confirmed by a quantitative cell-culture assay. J Gen Virol 87:697–704

Emerson SU, Purcell RH (2003) Hepatitis E virus. Rev Med Virol 13:145–154

Engle RE, Yu C, Emerson SU, Meng XJ, Purcell RH (2002) Hepatitis E virus (HEV) capsid antigens derived from viruses of human and swine origin are equally efficient for detecting anti-HEV by enzyme immunoassay. J Clin Microbiol 40:4576–4580

Erker JC, Desai SM, Mushahwar IK (1999) Rapid detection of Hepatitis E virus RNA by reverse transcription-polymerase chain reaction using universal oligonucleotide primers. J Virol Methods 81:109–113

Favorov MO, Fields HA, Purdy MA, Yashina TL, Aleksandrov AG, Alter MJ et al (1992) Serologic identification of hepatitis E virus infections in epidemic and endemic settings. J Med Virol 36:246–250

Favorov MO, Kosoy MY, Tsarev SA, Childs JE, Margolis HS (2000) Prevalence of antibody to hepatitis E virus among rodents in the United States. J Infect Dis 181:449–455

Feagins AR, Opriessnig T, Guenette DK, Halbur PG, Meng XJ (2007) Detection and characterization of infectious Hepatitis E virus from commercial pig livers sold in local grocery stores in the USA. J Gen Virol 88:912–917

Feagins AR, Opriessnig T, Guenette DK, Halbur PG, Meng XJ (2007) Detection and characterization of infectious Hepatitis E virus from commercial pig livers sold in local grocery stores in the USA. J Gen Virol 88:912–917

Feagins AR, Opriessnig T, Guenette DK, Halbur PG, Meng XJ (2008) Inactivation of infectious hepatitis E virus present in commercial pig livers sold in local grocery stores in the United States. Int J Food Microbiol 123:32–37

Fernandez-Barredo S, Galiana C, Garcia A, Gomez-Munoz MT (2007) Prevalence and genetic characterization of Hepatitis E virus in paired samples of feces and serum from naturally infected pigs. Can J Vet Res-Revue Canadienne De Recherche Veterinaire 71:236–240

Fernandez-Barredo S, Galiana C, Garcia A, Vega S, Gomez MT, Perez-Gracia MT (2006) Detection of hepatitis E virus shedding in feces of pigs at different stages of production using reverse transcription-polymerase chain reaction. J Vet Diagn Investig 18:462–5

Forgach P, Nowotny N, Erdelyi K, Boncz A, Zentai J, Szucs G et al (2010) Detection of hepatitis E virus in samples of animal origin collected in Hungary. Vet Microbiol 143:106–116

Galiana C, Fernandez-Barredo S, Garcia A, Gomez MT, Peerez-Gracia MT (2008) Short report: occupational exposure to hepatitis E virus (HEV) in swine workers. Am J Trop Med Hyg 78:1012–1015

Galiana C, Fernandez-Barredo S, Perez-Gracia MT (2010) Prevalence of hepatitis E virus (HEV) and risk factors in pig workers and blood donors. Enfermedades Infecciosas Y Microbiologia Clinica 28:602–607

Garbuglia AR, Scognamiglio P, Petrosillo N, Mastroianni CM, Sordillo P, Gentile D et al (2013) Hepatitis e virus genotype 4 outbreak, Italy, 2011. Emerg Infect Dis 19:110–114

Gardinali NR, Barry AF, da Silva PF, de Souza C, Alfieri AF, Alfieri AA (2012) Molecular detection and characterization of hepatitis E virus in naturally infected pigs from Brazilian herds. Res Vet Sci 93:1515–1519

Garkavenko O, Obriadina A, Meng J, Anderson DA, Benard HJ, Schroeder BA et al (2001) Detection and characterisation of swine hepatitis E virus in New Zealand. J Med Virol 65:525–529

Geng J, Fu H, Wang L, Bu Q, Liu P, Wang M et al (2011) Phylogenetic analysis of the full genome of rabbit hepatitis E virus (rbHEV) and molecular biologic study on the possibility of cross species transmission of rbHEV. Infect Genet Evol 11:2020–2025

Geng J, Wang L, Wang X, Fu H, Bu Q, Liu P et al (2011) Potential risk of zoonotic transmission from young swine to human: seroepidemiological and genetic characterization of hepatitis E virus in human and various animals in Beijing, China. J Viral Hepat 18:e583–e590

Geng J, Wang L, Wang X, Fu H, Bu Q, Zhu Y et al (2011) Study on prevalence and genotype of hepatitis E virus isolated from Rex Rabbits in Beijing, China. J Viral Hepat 18:661–667

Geng YS, Wang CB, Zhao CY, Yu XL, Harrison TJ, Tian KG et al (2010) Serological prevalence of Hepatitis E virus in domestic animals and diversity of Genotype 4 Hepatitis E virus in China. Vector-Borne Zoonotic Dis 10:765–770

Gessoni G, Manoni F (1996) Hepatitis E virus infection in north-east Italy: serological study in the open population and groups at risk. J Viral Hepat 3:197–202

Goens SD, Botero S, Hare W, Meng XJ, Perdue M (2003) Serological evidence for a hepatitis E virus of cattle. In: 22nd Annual Meeting of the American Society for Virology. Davis, pp 179

Goyal R, Kumar A, Panda SK, Paul SB, Acharya SK (2012) Ribavirin therapy for hepatitis E virus-induced acute on chronic liver failure: a preliminary report. Antivir Ther 17:1091–1096

Grandadam M, Tebbal S, Caron M, Siriwardana M, Larouze B, Koeck JL et al (2004) Evidence for hepatitis E virus quasispecies. J Gen Virol 85:3189–3194

Guerrero-Latorre L, Carratala A, Rodriguez-Manzano J, Calgua B, Hundesa A, Girones R (2011) Occurrence of water-borne enteric viruses in two settlements based in Eastern Chad: analysis of hepatitis E virus, hepatitis A virus and human adenovirus in water sources. J Water Health 9:515–524

Guo H, Zhou EM, Sun ZF, Meng XJ, Halbur PG (2006) Identification of B-cell epitopes in the capsid protein of avian hepatitis E virus (avian HEV) that are common to human and swine HEVs or unique to avian HEV. J Gen Virol 87:217–223

Guthmann JP, Klovstad H, Boccia D, Hamid N, Pinoges L, Nizou JY et al (2006) A large outbreak of hepatitis E among a displaced population in Darfur, Sudan, 2004: the role of water treatment methods. Clin Infect Dis 42:1685–1691

Gyarmati P, Mohammed N, Norder H, Blomberg J, Belak S, Widen F (2007) Universal detection of hepatitis E virus by two real-time PCR assays: TaqMan and Primer-probe energy transfer. J Virol Methods 146:226–235

Haim-Boukobza S, Ferey MP, Vetillard AL, Jeblaoui A, Pelissier E, Pelletier G et al (2012) Transfusion-transmitted hepatitis E in a misleading context of autoimmunity and drug-induced toxicity. J Hepatol 57:1374–1378

Hakze-van der Honing RW, van Coillie E, Antonis AF, van der Poel WH (2011) First isolation of hepatitis E virus genotype 4 in Europe through swine surveillance in the Netherlands and Belgium. PLoS One 6:e22673

Halac U, Beland K, Lapierre P, Patey N, Ward P, Brassard J et al (2012) Chronic hepatitis E infection in children with liver transplantation. Gut 61:597–603

Halbur PG, Kasorndorkbua C, Gilbert C, Guenette D, Potters MB, Purcell RH et al (2001) Comparative pathogenesis of infection of pigs with hepatitis E viruses recovered from a pig and a human. J Clin Microbiol 39:918–923

Halleux D, Kanaan N, Kabamba B, Thomas I, Hassoun Z (2012) Hepatitis E virus: an underdiagnosed cause of chronic hepatitis in renal transplant recipients. Transpl Infect Dis 14:99–102

Haqshenas G, Huang FF, Fenaux M, Guenette DK, Pierson FW, Larsen CT et al (2002) The putative capsid protein of the newly identified avian hepatitis E virus shares antigenic epitopes with that of swine and human hepatitis E viruses and chicken big liver and spleen disease virus. J Gen Virol 83:2201–2209

Haqshenas G, Shivaprasad HL, Woolcock PR, Read DH, Meng XJ (2001) Genetic identification and characterization of a novel virus related to human hepatitis E virus from chickens with hepatitis-splenomegaly syndrome in the United States. J Gen Virol 82:2449–2462

Heinz G, Hautzinger P (2007) Meat products with high levels of extenders and fillers. In: Meat processing technology for small- to medium-scale producers, pp 195–212. ftp://ftp.fao.org/do crep/fao/010/ai407e/aie07.pdf (Accessed 28 Feb 2013), Food and Agriculture Organization of the United Nations Regional Office For Asia and the Pacific, Bangkok

Hirano M, Ding X, Li TC, Takeda N, Kawabata H, Koizumi N et al (2003) Evidence for widespread infection of hepatitis E virus among wild rats in Japan. Hepatol Res 27:1–5

Hosmillo M, Jeong YJ, Kim HJ, Park JG, Nayak MK, Alfajaro MM et al (2010) Molecular detection of genotype 3 porcine hepatitis E virus in aborted fetuses and their sows. Arch Virol 155:1157–1161

Howard CM, Handzel T, Hill VR, Grytdal SP, Blanton C, Kamili S et al (2010) Novel risk factors associated with hepatitis E virus infection in a large outbreak in northern Uganda: results from a case-control study and environmental analysis. Am J Trop Med Hyg 83:1170–1173

Hsieh SY, Meng XJ, Wu YH, Liu ST, Tam AW, Lin DY et al (1999) Identity of a novel swine hepatitis E virus in Taiwan forming a monophyletic group with Taiwan isolates of human hepatitis E virus. J Clin Microbiol 37:3828–3834

Hu WP, Lu Y, Precioso NA, Chen HY, Howard T, Anderson D et al (2008) Double-antigen enzyme-linked immunosorbent assay for detection of hepatitis E virus-specific antibodies in human or swine sera. Clin Vaccine Immunol 15:1151–1157

Huang FF, Haqshenas G, Guenette DK, Halbur PG, Schommer SK, Pierson FW et al (2002) Detection by reverse transcription-PCR and genetic characterization of field isolates of swine hepatitis E virus from pigs in different geographic regions of the United States. J Clin Microbiol 40:1326–1332

Huang FF, Sun ZF, Emerson SU, Purcell RH, Shivaprasad HL, Pierson FW et al (2004) Determination and analysis of the complete genomic sequence of avian hepatitis E virus (avian HEV) and attempts to infect rhesus monkeys with avian HEV. J Gen Virol 85:1609–1618

Huang R, Li D, Wei S, Li Q, Yuan X, Geng L et al (1999) Cell culture of sporadic hepatitis E virus in China. Clin Diagn Lab Immunol 6:729–733

Huang S, Zhang X, Jiang H, Yan Q, Ai X, Wang Y et al (2010) Profile of acute infectious markers in sporadic hepatitis E. PLoS One 5:e13560

Innis BL, Seriwatana J, Robinson RA, Shrestha MP, Yarbough PO, Longer CF et al (2002) Quantitation of immunoglobulin to hepatitis E virus by enzyme immunoassay. Clin Diagn Lab Immunol 9:639–648

Ippagunta SK, Naik S, Sharma B, Aggarwal R (2007) Presence of hepatitis E virus in sewage in Northern India: frequency and seasonal pattern. J Med Virol 79:1827–1831

Ji Y, Zhu Y, Liang J, Wei X, Yang X, Wang L et al (2008) Swine hepatitis E virus in rural southern China: genetic characterization and experimental infection in rhesus monkeys (Macaca mulatta). J Gastroenterol 43:565–570

Jimenez de Oya N, de Blas I, Blazquez AB, Martin-Acebes MA, Halaihel N, Girones O et al (2011) Widespread distribution of hepatitis E virus in Spanish pig herds. BMC Res Notes 4:412

Jimenez de Oya N, Galindo I, Girones O, Duizer E, Escribano JM, Saiz JC (2009) Serological immunoassay for detection of hepatitis E virus on the basis of genotype 3 open reading frame 2 recombinant proteins produced in Trichoplusia ni larvae. J Clin Microbiol 47:3276–3282

Johne R, Dremsek P, Kindler E, Schielke A, Plenge-Bonig A, Gregersen H et al (2012) Rat hepatitis E virus: geographical clustering within Germany and serological detection in wild Norway rats (Rattus norvegicus). Infect Genet Evol 12:947–956

Johne R, Heckel G, Plenge-Bonig A, Kindler E, Maresch C, Reetz J et al (2010) Novel hepatitis E virus genotype in Norway rats, Germany. Emerg Infect Dis 16:1452–1455

Johne R, Plenge-Bonig A, Hess M, Ulrich RG, Reetz J, Schielke A (2010) Detection of a novel hepatitis E-like virus in faeces of wild rats using a nested broad-spectrum RT-PCR. J Gen Virol 91:750–758

Joshi MS, Walimbe AM, Arankalle VA, Chadha MS, Chitambar SD (2002) Hepatitis E antibody profiles in serum and urine. J Clin Lab Anal 16:137–142

Jothikumar N, Aparna K, Kamatchiammal S, Paulmurugan R, Saravanadevi S, Khanna P (1993) Detection of hepatitis E virus in raw and treated wastewater with the polymerase chain reaction. Appl Environ Microbiol 59:2558–2562

Jothikumar N, Cromeans TL, Robertson BH, Meng XJ, Hill VR (2006) A broadly reactive one-step real-time RT-PCR assay for rapid and sensitive detection of hepatitis E virus. J Virol Methods 131:65–71

Kaba M, Brouqui P, Richet H, Badiaga S, Gallian P, Raoult D et al (2010) Hepatitis E virus infection in sheltered homeless persons, France. Emerg Infect Dis 16:1761–1763

Kaba M, Colson P, Musongela JP, Tshilolo L, Davoust B (2010) Detection of hepatitis E virus of genotype 3 in a farm pig in Kinshasa (Democratic Republic of the Congo). Infect Genet Evol 10:154–157

Kaba M, Colson P, Musongela JP, Tshilolo L, Davoust B (2010) Detection of hepatitis E virus of genotype 3 in a farm pig in Kinshasa (Democratic Republic of the Congo). Infect Genet Evol 10:154–157

Kaba M, Davoust B, Cabre O, Colson P (2011) Hepatitis E virus genotype 3f in pigs in New Caledonia. Aust Vet J 89:496–499

Kaba M, Davoust B, Marie JL, Colson P (2010) Detection of hepatitis E virus in wild boar (Sus scrofa) livers. Vet J 186:259–261

Kabrane-Lazizi Y, Fine JB, Elm J, Glass GE, Higa H, Diwan A et al (1999) Evidence for widespread infection of wild rats with hepatitis E virus in the United States. Am J Trop Med Hyg 61:331–335

Kalia M, Chandra V, Rahman SA, Sehgal D, Jameel S (2009) Heparan sulfate proteoglycans are required for cellular binding of the hepatitis E virus ORF2 capsid protein and for viral infection. J Virol 83:12714–12724

Kamar N, Bendall R, Legrand-Abravanel F, Xia NS, Ijaz S, Izopet J et al (2012) Hepatitis E. Lancet 379:2477–2488

Kamar N, Bendall RP, Peron JM, Cintas P, Prudhomme L, Mansuy JM et al (2011) Hepatitis E virus and neurologic disorders. Emerg Infect Dis 17:173–179

Kanai Y, Miyasaka S, Uyama S, Kawami S, Kato-Mori Y, Tsujikawa M et al (2012) Hepatitis E virus in Norway rats (Rattus norvegicus) captured around a pig farm. BMC Res Notes 5:4

Kane MA, Bradley DW, Shrestha SM, Maynard JE, Cook EH, Mishra RP et al (1984) Epidemic non-A, non-B hepatitis in Nepal. Recovery of a possible etiologic agent and transmission studies in marmosets. JAMA 252:3140–3145

Kantala T, Maunula L, von Bonsdorff CH, Peltomaa J, Lappalainen M (2009) Hepatitis E virus in patients with unexplained hepatitis in Finland. J Clin Virol 45:109–113

Karetnyi YV, Gilchrist MJ, Naides SJ (1999) Hepatitis E virus infection prevalence among selected populations in Iowa. J Clin Virol 14:51–55

Karpe YA, Lole KS (2010) RNA 5'-triphosphatase activity of the hepatitis E virus helicase domain. J Virol 84:9637–9641

Kase JA, Correa MT, Luna C, Sobsey MD (2008) Isolation, detection and characterization of swine hepatitis E virus from herds in Costa Rica. Int J Environ Health Res 18:165–76

Kasorndorkbua C, Guenette DK, Huang FF, Thomas PJ, Meng XJ, Halbur PG (2004) Routes of transmission of swine hepatitis E virus in pigs. J Clin Microbiol 42:5047–5052

Kasorndorkbua C, Halbur PG, Thomas PJ, Guenette DK, Toth TE, Meng XJ (2002) Use of a swine bioassay and a RT-PCR assay to assess the risk of transmission of swine hepatitis E virus in pigs. J Virol Methods 101:71–78

Kasorndorkbua C, Opriessnig T, Huang FF, Guenette DK, Thomas PJ, Meng XJ et al (2005) Infectious swine hepatitis E virus is present in pig manure storage facilities on United States farms, but evidence of water contamination is lacking. Appl Environ Microbiol 71:7831–7837

Kasorndorkbua C, Thacker BJ, Halbur PG, Guenette DK, Buitenwerf RM, Royer RL et al (2003) Experimental infection of pregnant gilts with swine hepatitis E virus. Can J Vet Res 67:303–306

Khudyakov Y, Kamili S (2011) Serological diagnostics of hepatitis E virus infection. Virus Res 161:84–92

Khuroo MS (1991) Hepatitis E: the enterically transmitted non-A, non-B hepatitis. Indian J Gastroenterol 10:96–100

Khuroo MS, Kamili S, Khuroo MS (2009) Clinical course and duration of viremia in vertically transmitted hepatitis E virus (HEV) infection in babies born to HEV-infected mothers. J Viral Hepat 16:519–523

Khuroo MS, Kamili S, Yattoo GN (2004) Hepatitis E virus infection may be transmitted through blood transfusions in an endemic area. J Gastroenterol Hepatol 19:778–784

Kim SE, Kim MY, Kim DG, Song YJ, Jeong HJ, Lee SW et al (2008) Determination of fecal shedding rates and genotypes of swine hepatitis E virus (HEV) in Korea. J Vet Med Sci 70:1367–1371

Kokkinos P, Kozyra I, Lazic S, Bouwknegt M, Rutjes S, Willems K et al (2012) Harmonised investigation of the occurrence of human enteric viruses in the leafy green vegetable supply chain in three European countries. Food Environ Virol 4:179–191

Koning L, Pas SD, de Man RA, Balk AH, de Knegt RJ, ten Kate FJ et al (2013) Clinical implications of chronic hepatitis E virus infection in heart transplant recipients. J Heart Lung Transplant 32:78–85

Korkaya H, Jameel S, Gupta D, Tyagi S, Kumar R, Zafrullah M et al (2001) The ORF3 protein of hepatitis E virus binds to Src homology 3 domains and activates MAPK. J Biol Chem 276:42389–42400

Kulkarni MA, Arankalle VA (2008) The detection and characterization of hepatitis E virus in pig livers from retail markets of India. J Med Virol 80:1387–1390

Kuno A, Ido K, Isoda N, Satoh Y, Ono K, Satoh S et al (2003) Sporadic acute hepatitis E of a 47-year-old man whose pet cat was positive for antibody to hepatitis E virus. Hepatol Res 26:237–242

Lack JB, Volk K, Van Den Bussche RA (2012) Hepatitis E virus genotype 3 in wild rats, United States. Emerg Infect Dis 18:1268–1273

Lan X, Yang B, Li BY, Yin XP, Li XR, Liu JX (2009) Reverse transcription-loop-mediated isothermal amplification assay for rapid detection of hepatitis E virus. J Clin Microbiol 47:2304–2306

Leblanc D, Poitras E, Gagne MJ, Ward P, Houde A (2010) Hepatitis E virus load in swine organs and tissues at slaughterhouse determined by real-time RT-PCR. Int J Food Microbiol 139:206–209

Leblanc D, Ward P, Gagne MJ, Poitras E, Muller P, Trottier YL et al (2007) Presence of hepatitis E virus in a naturally infected swine herd from nursery to slaughter. Int J Food Microbiol 117:160–166

Lee SH, Kang SC, Kim DY, Bae JH, Kim JH (2007) Detection of swine hepatitis E virus in the porcine hepatic lesion in Jeju Island. J Vet Sci 8:51–55

Lee YH, Ha Y, Ahn KK, Chae C (2009) Localisation of swine hepatitis E virus in experimentally infected pigs. Vet J 179:417–421

Lee YH, Ha Y, Ahn KK, Cho KD, Lee BH, Kim SH et al (2009) Comparison of a new synthetic, peptide-derived, polyclonal antibody-based, immunohistochemical test with in situ hybridisation for the detection of swine hepatitis E virus in formalin-fixed, paraffin-embedded tissues. Vet J 182:131–135

Legrand-Abravanel F, Kamar N, Sandres-Saune K, Garrouste C, Dubois M, Mansuy JM et al (2010) Characteristics of autochthonous hepatitis E virus infection in solid-organ transplant recipients in France. J Infect Dis 202:835–844

Legrand-Abravanel F, Thevenet I, Mansuy JM, Saune K, Vischi F, Peron JM et al (2009) Good performance of immunoglobulin M assays in diagnosing genotype 3 hepatitis E virus infections. Clin Vaccine Immunol 16:772–774

Lhomme S, Abravanel F, Dubois M, Sandres-Saune K, Rostaing L, Kamar N et al (2012) Hepatitis E virus quasispecies and the outcome of acute hepatitis E in solid-organ transplant patients. J Virol 86:10006–10014

Li S, Tang X, Seetharaman J, Yang C, Gu Y, Zhang J et al (2009) Dimerization of hepatitis E virus capsid protein E2s domain is essential for virus-host interaction. PLoS Pathog 5:e1000537

Li SW, Zhang J, He ZQ, Gu Y, Liu RS, Lin J et al (2005) Mutational analysis of essential interactions involved in the assembly of hepatitis E virus capsid. J Biol Chem 280:3400–3406

Li TC, Ami Y, Suzaki Y, Yasuda SP, Yoshimatsu K, Arikawa J et al (2013) Characterization of full genome of rat hepatitis e virus strain from Vietnam. Emerg Infect Dis 19:115–118

Li TC, Chijiwa K, Sera N, Ishibashi T, Etoh Y, Shinohara Y et al (2005) Hepatitis E virus transmission from wild boar meat. Emerg Infect Dis 11:1958–1960

Li TC, Miyamura T, Takeda N (2007) Detection of hepatitis E virus RNA from the bivalve Yamato-Shijimi (Corbicula japonica) in Japan. Am J Trop Med Hyg 76:170–172

Li TC, Saito M, Ogura G, Ishibashi O, Miyamura T, Takeda N (2006) Serologic evidence for hepatitis E virus infection in mongoose. Am J Trop Med Hyg 74:932–936

Li TC, Yamakawa Y, Suzuki K, Tatsumi M, Razak MA, Uchida T et al (1997) Expression and self-assembly of empty virus-like particles of hepatitis E virus. J Virol 71:7207–7213

Li TC, Zhang J, Shinzawa H, Ishibashi M, Sata M, Mast EE et al (2000) Empty virus-like particle-based enzyme-linked immunosorbent assay for antibodies to hepatitis E virus. J Med Virol 62:327–333

Li W, Shu X, Pu Y, Bi J, Yang G, Yin G (2011) Seroprevalence and molecular detection of hepatitis E virus in Yunnan Province, China. Arch Virol 156:1989–1995

Liu JF, Zhang W, Shen Q, Yang SX, Huang F, Li PF et al (2009) Prevalence of antibody to hepatitis E virus among pet dogs in the Jiang-Zhe area of China. Scand J Infect Dis 41:291–295

Lorenzo FR, Tsatsralt-Od B, Ganbat S, Takahashi M, Okamoto H (2007) Analysis of the full-length genome of hepatitis E virus isolates obtained from farm pigs in Mongolia. J Med Virol 79:1128–1137

Lu L, Li C, Hagedorn CH (2006) Phylogenetic analysis of global hepatitis E virus sequences: genetic diversity, subtypes and zoonosis. Rev Med Virol 16:5–36

Ma H, Song X, Harrison TJ, Zhang H, Huang W, Wang Y (2011) Hepatitis E virus ORF3 antigens derived from genotype 1 and 4 viruses are detected with varying efficiencies by an anti-HEV enzyme immunoassay. J Med Virol 83:827–832

Ma H, Song X, Li Z, Harrison TJ, Zhang H, Huang W et al (2009) Varying abilities of recombinant polypeptides from different regions of hepatitis E virus ORF2 and ORF3 to detect anti-HEV immunoglobulin M. J Med Virol 81:1052–1061

Mansuy JM, Bendall R, Legrand-Abravanel F, Saune K, Miedouge M, Ellis V et al (2011) Hepatitis E virus antibodies in blood donors, France. Emerg Infect Dis 17:2309–2312

Mansuy JM, Peron JM, Abravanel F, Poirson H, Dubois M, Miedouge M et al (2004) Hepatitis E in the south west of France in individuals who have never visited an endemic area. J Med Virol 74:419–424

Marek A, Bilic I, Prokofieva I, Hess M (2010) Phylogenetic analysis of avian hepatitis E virus samples from European and Australian chicken flocks supports the existence of a different genus within the Hepeviridae comprising at least three different genotypes. Vet Microbiol 145:54–61

Martelli F, Caprioli A, Zengarini M, Marata A, Fiegna C, Di Bartolo I et al (2008) Detection of hepatitis E virus (HEV) in a demographic managed wild boar (Sus scrofa scrofa) population in Italy. Vet Microbiol 126:74–81

Martelli F, Toma S, Di Bartolo I, Caprioli A, Ruggeri FM, Lelli D et al (2010) Detection of Hepatitis E Virus (HEV) in Italian pigs displaying different pathological lesions. Res Vet Sci 88:492–496

Martin M, Segales J, Huang FF, Guenette DK, Mateu E, de Deus N et al (2007) Association of hepatitis E virus (HEV) and postweaning multisystemic wasting syndrome (PMWS) with lesions of hepatitis in pigs. Vet Microbiol 122:16–24

Martin-Serrano J, Zang T, Bieniasz PD (2001) HIV-1 and Ebola virus encode small peptide motifs that recruit Tsg101 to sites of particle assembly to facilitate egress. Nat Med 7:1313–1319

Masia G, Orru G, Liciardi M, Desogus G, Coppola RC, Murru V et al (2009) Evidence of hepatitis E virus (HEV) infection in human and pigs in Sardinia, Italy. J Prev Med Hyg 50:227–231

Masuda J, Yano K, Tamada Y, Takii Y, Ito M, Omagari K et al (2005) Acute hepatitis E of a man who consumed wild boar meat prior to the onset of illness in Nagasaki, Japan. Hepatol Res 31:178–183

Matsubayashi K, Kang JH, Sakata H, Takahashi K, Shindo M, Kato M et al (2008) A case of transfusion-transmitted hepatitis E caused by blood from a donor infected with hepatitis E virus via zoonotic food-borne route. Transfusion 48:1368–1375

Matsuda H, Okada K, Takahashi K, Mishiro S (2003) Severe hepatitis E virus infection after ingestion of uncooked liver from a wild boar. J Infect Dis 188:944

Matsuura Y, Suzuki M, Yoshimatsu K, Arikawa J, Takashima I, Yokoyama M et al (2007) Prevalence of antibody to hepatitis E virus among wild sika deer, Cervus nippon, in Japan. Arch Virol 152:1375–1381

McCreary C, Martelli F, Grierson S, Ostanello F, Nevel A, Banks M (2008) Excretion of hepatitis E virus by pigs of different ages and its presence in slurry stores in the United Kingdom. Vet Rec 163:261–265

Melenhorst WB, Gu YL, Jaspers WJ, Verhage AH (2007) Locally acquired hepatitis E in the Netherlands: associated with the consumption of raw pig meat? Scand J Infect Dis 39:454–456

Meng XJ (2011) From barnyard to food table: the omnipresence of hepatitis E virus and risk for zoonotic infection and food safety. Virus Res 161:23–30

Meng XJ (2010) Hepatitis E virus: animal reservoirs and zoonotic risk. Vet Microbiol 140:256–265

Meng XJ (2010) Recent advances in Hepatitis E virus. J Viral Hepat 17:153–161

Meng XJ (2003) Swine hepatitis E virus: cross-species infection and risk in xenotransplantation. Curr Top Microbiol Immunol 278:185–216

Meng XJ, Anderson DA, Arankalle VA, Emerson SU, Harrison TJ, Jameel S et al (2012) Hepeviridae. In: Virus taxonomy: ninth report of the international committee on the taxonomy of viruses. Academic Press, London, pp 1021–1028

Meng XJ, Halbur PG, Haynes JS, Tsareva TS, Bruna JD, Royer RL et al (1998) Experimental infection of pigs with the newly identified swine hepatitis E virus (swine HEV), but not with human strains of HEV. Arch Virol 143:1405–1415

Meng XJ, Halbur PG, Shapiro MS, Govindarajan S, Bruna JD, Mushahwar IK et al (1998) Genetic and experimental evidence for cross-species infection by swine hepatitis E virus. J Virol 72:9714–9721

Meng XJ, Purcell RH, Halbur PG, Lehman JR, Webb DM, Tsareva TS et al (1997) A novel virus in swine is closely related to the human hepatitis E virus. Proc Natl Acad Sci USA 94:9860–9865

Meng XJ, Shivaprasad HL, Payne C (2008) Hepatitis E virus infections. In: Saif YM, Fadly AM, Glisson JR, McDougald LR, Nolan LK, Swayne DE (eds) Diseases of poultry 12th edn. Blackwell Publishing Press, Ames, pp 443–452

Meng XJ, Wiseman B, Elvinger F, Guenette DK, Toth TE, Engle RE et al (2002) Prevalence of antibodies to hepatitis E virus in veterinarians working with swine and in normal blood donors in the United States and other countries. J Clin Microbiol 40:117–122

Mizuo H, Suzuki K, Takikawa Y, Sugai Y, Tokita H, Akahane Y et al (2002) Polyphyletic strains of hepatitis E virus are responsible for sporadic cases of acute hepatitis in Japan. J Clin Microbiol 40:3209–3218

Mizuo H, Yazaki Y, Sugawara K, Tsuda F, Takahashi M, Nishizawa T et al (2005) Possible risk factors for the transmission of hepatitis E virus and for the severe form of hepatitis E acquired locally in Hokkaido, Japan. J Med Virol 76:341–349

Moal V, Textoris J, Ben Amara A, Mehraj V, Berland Y, Colson P et al (2013) Chronic hepatitis E virus infection is specifically associated with an interferon-related transcriptional program. J Infect Dis 207:125–132

Morrow CJ, Samu G, Matrai E, Klausz A, Wood AM, Richter S et al (2008) Avian hepatitis E virus infection and possible associated clinical disease in broiler breeder flocks in Hungary. Avian Pathol 37:527–535

Munne MS, Vladimirsky S, Otegui L, Castro R, Brajterman L, Soto S et al (2006) Identification of the first strain of swine hepatitis E virus in South America and prevalence of anti-HEV antibodies in swine in Argentina. J Med Virol 78:1579–1583

Naik SR, Aggarwal R, Salunke PN, Mehrotra NN (1992) A large waterborne viral hepatitis E epidemic in Kanpur, India. Bull World Health Organ 70:597–604

Nakai I, Kato K, Miyazaki A, Yoshii M, Li TC, Takeda N et al (2006) Different fecal shedding patterns of two common strains of hepatitis E virus at three Japanese swine farms. Am J Trop Med Hyg 75:1171–1177

Nakamura M, Takahashi K, Taira K, Taira M, Ohno A, Sakugawa H et al (2006) Hepatitis E virus infection in wild mongooses of Okinawa, Japan: Demonstration of anti-HEV antibodies and a full-genome nucleotide sequence. Hepatol Res 34:137–140

Neuvonen M, Ahola T (2009) Differential activities of cellular and viral macro domain proteins in binding of ADP-ribose metabolites. J Mol Biol 385:212–225

Nidaira M, Takahashi K, Ogura G, Taira K, Okano S, Kudaka J et al (2012) Detection and phylogenetic analysis of hepatitis e viruses from mongooses in okinawa, Japan. J Vet Med Sci 74:1665–1668

Niikura M, Takamura S, Kim G, Kawai S, Saijo M, Morikawa S et al (2002) Chimeric recombinant hepatitis E virus-like particles as an oral vaccine vehicle presenting foreign epitopes. Virology 293:273–280

Nishizawa T, Takahashi M, Endo K, Fujiwara S, Sakuma N, Kawazuma F et al (2005) Analysis of the full-length genome of hepatitis E virus isolates obtained from wild boars in Japan. J Gen Virol 86:3321–3326

Nishizawa T, Takahashi M, Mizuo H, Miyajima H, Gotanda Y, Okamoto H (2003) Characterization of Japanese swine and human hepatitis E virus isolates of genotype IV with 99 % identity over the entire genome. J Gen Virol 84:1245–1251

Obriadina A, Meng JH, Ulanova T, Trinta K, Burkov A, Fields HA et al (2002) A new enzyme immunoassay for the detection of antibody to hepatitis E virus. J Gastroenterol Hepatol 17(Suppl 3):S360–S364

Okamoto H (2013) Culture systems for hepatitis E virus. J Gastroenterol 48:147–158

Okamoto H (2007) Genetic variability and evolution of hepatitis E virus. Virus Res 127:216–228

Okamoto H (2011) Hepatitis E virus cell culture models. Virus Res 161:65–77

Panda SK, Thakral D, Rehman S (2007) Hepatitis E virus. Rev Med Virol 17:151–180

Parashar D, Khalkar P, Arankalle VA (2011) Survival of hepatitis A and E viruses in soil samples. Clin Microbiol Infect 17:E1–E4

Pas SD, de Man RA, Mulders C, Balk AH, van Hal PT, Weimar W et al (2012) Hepatitis E virus infection among solid organ transplant recipients, the Netherlands. Emerg Infect Dis 18:869–872

Patra S, Kumar A, Trivedi SS, Puri M, Sarin SK (2007) Maternal and fetal outcomes in pregnant women with acute hepatitis E virus infection. Ann Intern Med 147:28–33

Pavio N, Meng XJ, Renou C (2010) Zoonotic hepatitis E: animal reservoirs and emerging risks. Vet Res 41:46

Payne CJ, Ellis TM, Plant SL, Gregory AR, Wilcox GE (1999) Sequence data suggests big liver and spleen disease virus (BLSV) is genetically related to hepatitis E virus. Vet Microbiol 68:119–125

Pei Y, Yoo D (2002) Genetic characterization and sequence heterogeneity of a canadian isolate of Swine hepatitis E virus. J Clin Microbiol 40:4021–4029

Pensaert MB, Sanchez RE Jr, Ladekjaer-Mikkelsen AS, Allan GM, Nauwynck HJ (2004) Viremia and effect of fetal infection with porcine viruses with special reference to porcine circovirus 2 infection. Vet Microbiol 98:175–183

Peralta B, Biarnes M, Ordonez G, Porta R, Martin M, Mateu E et al (2009) Evidence of widespread infection of avian hepatitis E virus (avian HEV) in chickens from Spain. Vet Microbiol 137:31–36

Peralta B, Casas M, de Deus N, Martin M, Ortuno A, Perez-Martin E et al (2009) Anti-HEV antibodies in domestic animal species and rodents from Spain using a genotype 3-based ELISA. Vet Microbiol 137:66–73

Pichlmair A, Schulz O, Tan CP, Naslund TI, Liljestrom P, Weber F et al (2006) RIG-I-mediated antiviral responses to single-stranded RNA bearing 5'-phosphates. Science 314:997–1001

Pina S, Buti M, Cotrina M, Piella J, Girones R (2000) HEV identified in serum from humans with acute hepatitis and in sewage of animal origin in Spain. J Hepatol 33:826–833

Pina S, Jofre J, Emerson SU, Purcell RH, Girones R (1998) Characterization of a strain of infectious hepatitis E virus isolated from sewage in an area where hepatitis E is not endemic. Appl Environ Microbiol 64:4485–4488

Prabhu SB, Gupta P, Durgapal H, Rath S, Gupta SD, Acharya SK et al (2011) Study of cellular immune response against Hepatitis E virus (HEV). J Viral Hepat 18:587–594

Purcell RH, Emerson SU (2010) Hidden danger: the raw facts about hepatitis E virus. J Infect Dis 202:819–821

Purcell RH, Engle RE, Rood MP, Kabrane-Lazizi Y, Nguyen HT, Govindarajan S et al (2011) Hepatitis E virus in rats, Los Angeles, California, USA. Emerg Infect Dis 17:2216–2222

Rab MA, Bile MK, Mubarik MM, Asghar H, Sami Z, Siddiqi S et al (1997) Water-borne hepatitis E virus epidemic in Islamabad, Pakistan: a common source outbreak traced to the malfunction of a modern water treatment plant. Am J Trop Med Hyg 57:151–157

Raj VS, Smits SL, Pas SD, Provacia LB, Moorman-Roest H, Osterhaus AD et al (2012) Novel hepatitis E virus in ferrets, the Netherlands. Emerg Infect Dis 18:1369–1370

Razonable RR (2011) Rare, unusual, and less common virus infections after organ transplantation. Curr Opin Organ Transplant 16:580–587

Razonable RR, Findlay JY, O'Riordan A, Burroughs SG, Ghobrial RM, Agarwal B et al (2011) Critical care issues in patients after liver transplantation. Liver Transpl 17:511–527

Rehman S, Kapur N, Durgapal H, Panda SK (2008) Subcellular localization of hepatitis E virus (HEV) replicase. Virology 370:77–92

Renou C, Moreau X, Pariente A, Cadranel JF, Maringe E, Morin T et al (2008) A national survey of acute hepatitis E in France. Aliment Pharmacol Ther 27:1086–1093

Renou C, Nicand E, Pariente A, Cadranel JF, Pavio N (2009) How to investigate and diagnose autochthonous hepatitis E? Gastroenterol Clin Biol 33:F27–F35

Reuter G, Fodor D, Forgach P, Katai A, Szucs G (2009) Characterization and zoonotic potential of endemic hepatitis E virus (HEV) strains in humans and animals in Hungary. J Clin Virol 44:277–281

Reuter G, Fodor D, Katai A, Szucs G (2006) Identification of a novel variant of human hepatitis E virus in Hungary. J Clin Virol 36:100–102

Ritchie SJ, Riddell C (1991) British Columbia. "Hepatitis-splenomegaly" syndrome in commercial egg laying hens. Can Vet J 32:500–501

Rogee S, Talbot N, Caperna T, Bouquet J, Barnaud E, Pavio N (2012) New models of hepatitis E virus replication in human and porcine hepatocyte cell lines. J Gen Virol

Romano L, Paladini S, Tagliacarne C, Canuti M, Bianchi S, Zanetti AR (2011) Hepatitis E in Italy: a long-term prospective study. J Hepatol 54:34–40

Rose N, Lunazzi A, Dorenlor V, Merbah T, Eono F, Eloit M et al (2011) High prevalence of Hepatitis E virus in French domestic pigs. Comp Immunol Microbiol Infect Dis 34:419–427

Rozanov MN, Koonin EV, Gorbalenya AE (1992) Conservation of the putative methyltransferase domain: a hallmark of the 'Sindbis-like' supergroup of positive-strand RNA viruses. J Gen Virol 73(Pt 8):2129–3134

Rutjes SA, Lodder WJ, Lodder-Verschoor F, van den Berg HH, Vennema H, Duizer E et al (2009) Sources of hepatitis E virus genotype 3 in The Netherlands. Emerg Infect Dis 15:381–387

Rutjes SA, Lodder-Verschoor F, Lodder WJ, van der Giessen J, Reesink H, Bouwknegt M et al (2010) Seroprevalence and molecular detection of hepatitis E virus in wild boar and red deer in The Netherlands. J Virol Methods 168:197–206

Saad MD, Hussein HA, Bashandy MM, Kamel HH, Earhart KC, Fryauff DJ et al (2007) Hepatitis E virus infection in work horses in Egypt. Infect Genet Evol 7:368–373

Said B, Ijaz S, Kafatos G, Booth L, Thomas HL, Walsh A et al (2009) Hepatitis E outbreak on cruise ship. Emerg Infect Dis 15:1738–1744

Sailaja B, Murhekar MV, Hutin YJ, Kuruva S, Murthy SP, Reddy KS et al (2009) Outbreak of waterborne hepatitis E in Hyderabad, India, 2005. Epidemiol Infect 137:234–240

Sanford BJ, Dryman BA, Huang YW, Feagins AR, Leroith T, Meng XJ (2011) Prior infection of pigs with a genotype 3 swine hepatitis E virus (HEV) protects against subsequent challenges with homologous and heterologous genotypes 3 and 4 human HEV. Virus Res 159:17–22

Sato Y, Sato H, Naka K, Furuya S, Tsukiji H, Kitagawa K et al (2011) A nationwide survey of hepatitis E virus (HEV) infection in wild boars in Japan: identification of boar HEV strains of genotypes 3 and 4 and unrecognized genotypes. Arch Virol 156:1345–1358

Satou K, Nishiura H (2007) Transmission dynamics of hepatitis E among swine: potential impact upon human infection. BMC Vet Res 3:9

Savic B, Milicevic V, Bojkovski J, Kureljusic B, Ivetic V, Pavlovic I (2010) Detection rates of the swine torque teno viruses (TTVs), porcine circovirus type 2 (PCV2) and hepatitis E virus (HEV) in the livers of pigs with hepatitis. Vet Res Commun 34:641–648

Schielke A, Sachs K, Lierz M, Appel B, Jansen A, Johne R (2009) Detection of hepatitis E virus in wild boars of rural and urban regions in Germany and whole genome characterization of an endemic strain. Virol J 6:58

Schlauder GG, Desai SM, Zanetti AR, Tassopoulos NC, Mushahwar IK (1999) Novel hepatitis E virus (HEV) isolates from Europe: evidence for additional genotypes of HEV. J Med Virol 57:243–251

Scobie L, Dalton HR (2013) Hepatitis E: source and route of infection, clinical manifestations and new developments. J Viral Hepat 20:1–11

Scotto G, Martinelli D, Giammario A, Prato R, Fazio V (2013) Prevalence of antibodies to hepatitis E virus in immigrants: a seroepidemiological survey in the district of Foggia (Apulia-southern Italy). J Med Virol 85:261–265

Segales J, Allan GM, Domingo M (2005) Porcine circovirus diseases. Anim Health Res Rev 6:119–142

Shata MT, Barrett A, Shire NJ, Abdelwahab SF, Sobhy M, Daef E et al (2007) Characterization of hepatitis E-specific cell-mediated immune response using IFN-gamma ELISPOT assay. J Immunol Methods 328:152–161

Shukla P, Chauhan UK, Naik S, Anderson D, Aggarwal R (2007) Hepatitis E virus infection among animals in northern India: an unlikely source of human disease. J Viral Hepat 14:310–317

Siripanyaphinyo U, Laohasinnarong D, Siripanee J, Kaeoket K, Kameoka M, Ikuta K et al (2009) Full-length sequence of genotype 3 hepatitis E virus derived from a pig in Thailand. J Med Virol 81:657–664

Song YJ, Jeong HJ, Kim YJ, Lee SW, Lee JB, Park SY et al (2010) Analysis of complete genome sequences of Swine Hepatitis E virus and possible risk factors for transmission of HEV to humans in Korea. J Med Virol 82:583–591

Sonoda H, Abe M, Sugimoto T, Sato Y, Bando M, Fukui E et al (2004) Prevalence of hepatitis E virus (HEV) Infection in wild boars and deer and genetic identification of a genotype 3 HEV from a boar in Japan. J Clin Microbiol 42:5371–5374

Srivastava R, Aggarwal R, Bhagat MR, Chowdhury A, Naik S (2008) Alterations in natural killer cells and natural killer T cells during acute viral hepatitis E. J Viral Hepat 15:910–916

Srivastava R, Aggarwal R, Jameel S, Puri P, Gupta VK, Ramesh VS et al (2007) Cellular immune responses in acute hepatitis E virus infection to the viral open reading frame 2 protein. Viral Immunol 20:56–65

Sun ZF, Larsen CT, Huang FF, Billam P, Pierson FW, Toth TE et al (2004) Generation and infectivity titration of an infectious stock of avian hepatitis E virus (HEV) in chickens and cross-species infection of turkeys with avian HEV. J Clin Microbiol 42:2658–2662

Surjit M, Oberoi R, Kumar R, Lal SK (2006) Enhanced alpha1 microglobulin secretion from Hepatitis E virus ORF3-expressing human hepatoma cells is mediated by the tumor susceptibility gene 101. J Biol Chem 281:8135–8142

Takahashi H, Tanaka T, Jirintai S, Nagashima S, Takahashi M, Nishizawa T et al (2012) A549 and PLC/PRF/5 cells can support the efficient propagation of swine and wild boar hepatitis E virus (HEV) strains: demonstration of HEV infectivity of porcine liver sold as food. Arch Virol 157:235–246

Takahashi K, Kitajima N, Abe N, Mishiro S (2004) Complete or near-complete nucleotide sequences of hepatitis E virus genome recovered from a wild boar, a deer, and four patients who ate the deer. Virology 330:501–505

Takahashi M, Kusakai S, Mizuo H, Suzuki K, Fujimura K, Masuko K et al (2005) Simultaneous detection of immunoglobulin A (IgA) and IgM antibodies against hepatitis E virus (HEV) Is highly specific for diagnosis of acute HEV infection. J Clin Microbiol 43:49–56

Takahashi M, Nishizawa T, Miyajima H, Gotanda Y, Iita T, Tsuda F et al (2003) Swine hepatitis E virus strains in Japan form four phylogenetic clusters comparable with those of Japanese isolates of human hepatitis E virus. J Gen Virol 84:851–862

Takahashi M, Nishizawa T, Sato H, Sato Y, Jirintai NS et al (2011) Analysis of the full-length genome of a hepatitis E virus isolate obtained from a wild boar in Japan that is classifiable into a novel genotype. J Gen Virol 92:902–908

Takamura S, Niikura M, Li TC, Takeda N, Kusagawa S, Takebe Y et al (2004) DNA vaccine-encapsulated virus-like particles derived from an orally transmissible virus stimulate mucosal and systemic immune responses by oral administration. Gene Ther 11:628–635

Tamada Y, Yano K, Yatsuhashi H, Inoue O, Mawatari F, Ishibashi H (2004) Consumption of wild boar linked to cases of hepatitis E. J Hepatol 40:869–870

Tei S, Kitajima N, Takahashi K, Mishiro S (2003) Zoonotic transmission of hepatitis E virus from deer to human beings. Lancet 362:371–373

Temmam S, Besnard L, Andriamandimby SF, Foray C, Rasamoelina-Andriamanivo H, Heraud JM et al (2013) High Prevalence of Hepatitis E in Humans and Pigs and Evidence of Genotype-3 Virus in Swine, Madagascar. Am J Trop Med Hyg 88:329–338

Teshale EH, Howard CM, Grytdal SP, Handzel TR, Barry V, Kamili S et al (2010) Hepatitis E epidemic, Uganda. Emerg Infect Dis 16:126–129

Tesse S, Lioure B, Fornecker L, Wendling MJ, Stoll-Keller F, Bigaillon C et al (2012) Circulation of genotype 4 hepatitis E virus in Europe: first autochthonous hepatitis E infection in France. J Clin Virol 54:197–200

Thomas DL, Yarbough PO, Vlahov D, Tsarev SA, Nelson KE, Saah AJ et al (1997) Seroreactivity to hepatitis E virus in areas where the disease is not endemic. J Clin Microbiol 35:1244–1247

Tomiyama D, Inoue E, Osawa Y, Okazaki K (2009) Serological evidence of infection with hepatitis E virus among wild Yezo-deer, Cervus nippon yesoensis, in Hokkaido, Japan. J Viral Hepat 16:524–528

Tsang TH, Denison EK, Williams HV, Venczel LV, Ginsberg MM, Vugia DJ (2000) Acute hepatitis E infection acquired in California. Clin Infect Dis 30:618–619

Tsarev SA, Tsareva TS, Emerson SU, Rippy MK, Zack P, Shapiro M et al (1995) Experimental Hepatitis-E in Pregnant Rhesus-Monkeys—failure to transmit Hepatitis-E Virus (Hev) to offspring and evidence of naturally acquired antibodies to Hev. J Infect Dis 172:31–37

Tsarev SA, Tsareva TS, Emerson SU, Yarbough PO, Legters LJ, Moskal T et al (1994) Infectivity titration of a prototype strain of hepatitis E virus in cynomolgus monkeys. J Med Virol 43:135–142

Utsumi T, Hayashi Y, Lusida MI, Amin M, Soetjipto Hendra A et al (2011) Prevalence of hepatitis E virus among swine and humans in two different ethnic communities in Indonesia. Arch Virol 156:689–693

Vaidya SR, Tilekar BN, Walimbe AM, Arankalle VA (2003) Increased risk of hepatitis E in sewage workers from India. J Occup Environ Med 45:1167–1170

van der Poel WH, Verschoor F, van der Heide R, Herrera MI, Vivo A, Kooreman M et al (2001) Hepatitis E virus sequences in swine related to sequences in humans, The Netherlands. Emerg Infect Dis 7:970–976

Vasickova P, Psikal I, Kralik P, Widen F, Hubalek Z, Pavlik I (2007) Hepatitis E virus: a review. Veterinarni Medicina 52:365–384

Wacheck S, Sarno E, Martlbauer E, Zweifel C, Stephan R (2012) Seroprevalence of anti-hepatitis E virus and anti-Salmonella antibodies in pigs at slaughter in Switzerland. J Food Prot 75:1483–1485

Wang H, He Y, Shen Q, Wang X, Yang S, Cui L et al (2012) Complete genome sequence of the genotype 4 hepatitis E virus strain prevalent in swine in Jiangsu Province, China, reveals a close relationship with that from the human population in this area. J Virol 86:8334–8335

Wang L, Zhuang H (2004) Hepatitis E: an overview and recent advances in vaccine research. World J Gastroenterol 10:2157–2162

Wang Y, Ma X (2010) Detection and sequences analysis of sheep hepatitis E virus RNA in Xinjiang autonomous region. Wei Sheng Wu Xue Bao 50:937–941

Wang Y, Zhang H, Li Z, Gu W, Lan H, Hao W et al (2001) Detection of sporadic cases of hepatitis E virus (HEV) infection in China using immunoassays based on recombinant open reading frame 2 and 3 polypeptides from HEV genotype 4. J Clin Microbiol 39:4370–4379

Wang YC, Zhang HY, Xia NS, Peng G, Lan HY, Zhuang H et al (2002) Prevalence, isolation, and partial sequence analysis of hepatitis E virus from domestic animals in China. J Med Virol 67:516–521

Ward P, Muller P, Letellier A, Quessy S, Simard C, Trottier YL et al (2008) Molecular characterization of hepatitis E virus detected in swine farms in the province of Quebec. Can J Vet Res 72:27–31

Watanobe T, Okumura N, Ishiguro N, Nakano M, Matsui A, Sahara M et al (1999) Genetic relationship and distribution of the Japanese wild boar (Sus scrofa leucomystax) and Ryukyu wild boar (Sus scrofa riukiuanus) analysed by mitochondrial DNA. Mol Ecol 8:1509–1512

Wenzel JJ, Preiss J, Schemmerer M, Huber B, Plentz A, Jilg W (2011) Detection of hepatitis E virus (HEV) from porcine livers in Southeastern Germany and high sequence homology to human HEV isolates. J Clin Virol 52:50–54

Wibawa ID, Muljono DH, Mulyanto, Suryadarma IG, Tsuda F, Takahashi M et al (2004) Prevalence of antibodies to hepatitis E virus among apparently healthy humans and pigs in Bali, Indonesia: Identification of a pig infected with a genotype 4 hepatitis E virus. J Med Virol 73:38–44

Wichmann O, Schimanski S, Koch J, Kohler M, Rothe C, Plentz A et al (2008) Phylogenetic and case-control study on hepatitis E virus infection in Germany. J Infect Dis 198:1732–1741

Widen F, Sundqvist L, Matyi-Toth A, Metreveli G, Belak S, Hallgren G et al (2011) Molecular epidemiology of hepatitis E virus in humans, pigs and wild boars in Sweden. Epidemiol Infect 139:361–371

Williams TPE, Kasorndorkbua C, Halbur PG, Haqshenas G, Guenette DK, Toth TE et al (2001) Evidence of extrahepatic sites of replication of the hepatitis E virus in a swine model. J Clin Microbiol 39:3040–3046

Withers MR, Correa MT, Morrow M, Stebbins ME, Seriwatana J, Webster WD et al (2002) Antibody levels to hepatitis E virus in North Carolina swine workers, non-swine workers, swine, and murids. Am J Trop Med Hyg 66:384–8

Wu JC, Chen CM, Chiang TY, Tsai WH, Jeng WJ, Sheen IJ et al (2002) Spread of hepatitis E virus among different-aged pigs: two-year survey in Taiwan. J Med Virol 66:488–492

Xing L, Wang JC, Li TC, Yasutomi Y, Lara J, Khudyakov Y et al (2011) Spatial configuration of hepatitis E virus antigenic domain. J Virol 85:1117–1124

Yamada K, Takahashi M, Hoshino Y, Takahashi H, Ichiyama K, Nagashima S et al (2009) ORF3 protein of hepatitis E virus is essential for virion release from infected cells. J Gen Virol 90:1880–1891

Yamamoto H, Suzuki J, Matsuda A, Ishida T, Ami Y, Suzaki Y et al (2012) Hepatitis E virus outbreak in monkey facility, Japan. Emerg Infect Dis 18:2032–2034

Yamashita T, Mori Y, Miyazaki N, Cheng RH, Yoshimura M, Unno H et al (2009) Biological and immunological characteristics of hepatitis E virus-like particles based on the crystal structure. Proc Natl Acad Sci USA 106:12986–12991

Yazaki Y, Mizuo H, Takahashi M, Nishizawa T, Sasaki N, Gotanda Y et al (2003) Sporadic acute or fulminant hepatitis E in Hokkaido, Japan, may be food-borne, as suggested by the presence of hepatitis E virus in pig liver as food. J Gen Virol 84:2351–2357

Yoo D, Willson P, Pei Y, Hayes MA, Deckert A, Dewey CE et al (2001) Prevalence of hepatitis E virus antibodies in Canadian swine herds and identification of a novel variant of swine hepatitis E virus. Clin Diagn Lab Immunol 8:1213–1219

Yu Y, Sun J, Liu M, Xia L, Zhao C, Harrison TJ et al (2009) Seroepidemiology and genetic characterization of hepatitis E virus in the northeast of China. Infect Genet Evol 9:554–561

Zanetti AR, Dawson GJ (1994) Hepatitis type E in Italy: a seroepidemiological survey. Study group of Hepatitis E. J Med Virol 42:318–320

Zhai L, Dai X, Meng J (2006) Hepatitis E virus genotyping based on full-length genome and partial genomic regions. Virus Res 120:57–69

Zhang H, Mohn U, Prickett JR, Schalk S, Motz M, Halbur PG et al (2011) Differences in capabilities of different enzyme immunoassays to detect anti-hepatitis E virus immunoglobulin G in pigs infected experimentally with hepatitis E virus genotype 3 or 4 and in pigs with unknown exposure. J Virol Methods 175:156–162

Zhang LQ, Zhao FR, Liu ZG, Kong WL, Wang H, Ouyang Y et al (2012) Simple and rapid detection of swine hepatitis E virus by reverse transcription loop-mediated isothermal amplification. Arch Virol 157:2383–2388

Zhang W, Shen Q, Mou J, Gong G, Yang Z, Cui L et al (2008) Hepatitis E virus infection among domestic animals in eastern China. Zoonoses Public Health 55:291–298

Zhao C, Ma Z, Harrison TJ, Feng R, Zhang C, Qiao Z et al (2009) A novel genotype of hepatitis E virus prevalent among farmed rabbits in China. J Med Virol 81:1371–1379

Zhao Q, Zhang J, Wu T, Li SW, Ng MH, Xia NS et al (2012) Antigenic determinants of hepatitis E virus and vaccine-induced immunogenicity and efficacy. J Gastroenterol

Zheng Y, Ge S, Zhang J, Guo Q, Ng MH, Wang F et al (2006) Swine as a principal reservoir of hepatitis E virus that infects humans in eastern China. J Infect Dis 193:1643–1649

Zhu FC, Zhang J, Zhang XF, Zhou C, Wang ZZ, Huang SJ et al (2010) Efficacy and safety of a recombinant hepatitis E vaccine in healthy adults: a large-scale, randomised, double-blind placebo-controlled, phase 3 trial. Lancet 376:895–902

Zhuang H, Cao XY, Liu CB, Wang GM (1991) Epidemiology of hepatitis E in China. Gastroenterol Jpn 26(Suppl 3):135–138

Zwettler D, Fink M, Revilla-Fernandez S, Steinrigl A, Winter P, Kofer J (2012) First detection of hepatitis E virus in Austrian pigs by RT-qPCR. Berl Munch Tierarztl Wochenschr 125:281–289

Index

F. M. Ruggeri et al., *Hepatitis E Virus*, SpringerBriefs in Food, Health, and Nutrition,   87
DOI: 10.1007/978-1-4614-7522-4, © Franco Maria Ruggeri, Ilaria Di Bartolo,
Marcello Trevisani, Fabio Ostanello 2013